Biotechnology Science: Tools and Techniques

Biotechnology Science: Tools and Techniques

P. R. Gilbert

ANMOL PUBLICATIONS PVT. LTD.
NEW DELHI - 110 002 (INDIA)

ANMOL PUBLICATIONS PVT. LTD.
H.O.: 4374/4B, Ansari Road, Darya Ganj
New Delhi-110 002 (India)
Ph.: 23278000, 23261597
B.O.: 1015, Ist Main Road, BSK IIIrd Stage
IIIrd Phase, IIIrd Block
Bangalore - 560 085 (Karnataka)
Ph: 080-41723429, Tel/Fax: 080-2672 3604
Visit us at: www.anmolpublications.com

Biotechnology Science: Tools and Techniques

First Published, 2008

ISBN 978-81-261-3281-2

PRINTED IN INDIA

Printed at Mehra Offset Press, Delhi.

Contents

Preface

This book entitled "Biotechnology Science, Tools and Techniques" is designed to acquaint readers with the modern biotechnology science, tools and techniques. Biotechnology has many applications in health, agriculture, aquaculture and the environment. Health Canada's role in biotechnology is to protect the health and safety of Canadians through regulation and public health policy. Humans have used biotechnology for thousands of years. For example, using microorganisms such as bacteria or fungi has used biotechnology to make cheese, ferment wine and beer and make bread. Over time, we have also domesticated and selectively bred some animals and plants to meet human needs. In Canada, Health Canada, the Canadian Food Inspection Agency, Fisheries and Oceans Canada and Environment Canada, have joint responsibility to regulate biotechnology-derived products.

The definition used is found in the Canadian Environmental Protection Act (1999): "the application of science and engineering to the direct or indirect use of living organisms of parts or products of living organisms, in their natural or modified forms." Examples of biotechnology-derived products regulated by Health Canada are genetically modified and other novel foods, biologics and genetic therapies and assisted human reproduction technologies. Biotechnology can be best understood as an umbrella term referring to a variety of techniques, both traditional and modern. Note that some traditional techniques, such as selective breeding, hybridization and mutagenesis, are used in current applications of biotechnology, only with increased scientific knowledge and more advanced technology.

These include fermentation (the breakdown of complex organic substances by micro-organisms, in the absence of oxygen; the process is energy-yielding), selective breeding (the breeding of selected plants and animals to produce offspring with desired traits; the offspring with the desired traits are then used as breeding stock for the next generation and so on, until offspring that express the desired traits are obtained), hybridization (the production of offspring, known as hybrids, from genetically dissimilar parents; the object of hybridization is to combine desirable genes found in two or more different varieties to produce pure-breeding offspring superior in many respects to the parental types); mutagenesis (the use of mutagens ,such as exposure to radiation, temperature extremes and certain chemicals, to cause changes in the genetic make-up of cells, possibly resulting in new desirable, inheritable traits); recombinant DNA (RDNA) techniques (the application of genetic techniques to produce desirable traits in living organisms using other living organisms such as bacteria. Examples of these techniques include the use of: restriction endonucleases (enzymes, produced by bacteria, that break foreign DNA molecules with the gene of interest into fragments which recombine with complementary molecules from a different source to form a recombinant DNA molecule); vectors (plasmids, often bacterial, or viruses that carry a piece of DNA into a bacterium for cloning purposes); and gene guns (called DNA particle guns and used to fire genes into cells by coating the genes onto tiny gold or tungsten particles and firing them from the barrel of the gun); tissue culture (a biological technique in which fragments of plant or animal tissue or cells are transferred to an artificial environment, free of other organisms, in which they continue to survive and function for reproduction, chemical production and medical research).

This book consists of—Biotechnology Science: An Introductory Overviewof Basics; Modern Advances in Biotechnology Science; Future Prospects of Biotechnology Science; and Biotech Science: Tools and Techniques etc. besides a large body of list of acronyms, glossary of relevant terms, extensive bibliography, list of websites and links, for further referencing and research.

—Editor

List of Acronyms

AACC	—	American Association for Clinical Chemistry
AACR	—	American Association for Cancer Research
AAI	—	American Association of Immunologists
AAIP	—	American Association of Investigative Pathologists
AAP	—	Association of American Physicians
ABA	—	Australian Biotechnol. Assoc.
ABI	—	Applied Biosystems, Inc.
ABIC	—	Agricultural Biotechnology Intl. Conf.
ABRF	—	Association of Biomolecular Resource Facilities
ACMG	—	American College of Medical Genet.
ACS	—	American Chemical Society
ADA	—	Americans with Disabilities Act
AEC	—	Atomic Energy Commission
AES	—	American Electrophoresis Society
AFMR	—	American Federation of Medical Research
AFIP/ARP	—	Armed Forces Inst. of Pathol./Am. Registry of Pathol.
AGT	—	Association of Genetic Technologists

AHA	—	American Heart Association
AIBS	—	Am. Inst. of Biol. Soc.
AMIA	—	American Medical Informatics Association
ANGIS	—	Australian National Genomic Info. Service
ANL	—	Argonne National Laboratory
API	—	Application Programming Interface
ARP	—	American Registry of Pathology
ASB	—	American Society for Biotechnology
ASBMB	—	American Society for Biochemistry & Molecular Biology
ASCB	—	American Society for Cell Biology
ASCI	—	American Society for Clinical Investigation
ASHG	—	American Society for Human Genetics
ASIP	—	American Society for Investigative Pathologists
ASIS	—	American Society for Information Science
ASLME	—	American Society of Law, Medicine, and Ethics
ASM	—	American Society for Microbiology
ASPET	—	American Society for Pharmacology and Experimental Therapeutics
ATCC	—	American Type Culture Collection
ATP	—	Advanced Technology Program
AVS	—	American Vacuum Society
AWCH	—	Adelaide Women and Children's Hospital

BAC	—	Bacterial Artificial Chromosome
BACR	—	British Association for Cancer Research
BCM	—	Baylor College of Medicine
BDG	—	Batten's disease gene
BER	—	Biological and Environmental Research
BIO	—	Biotechnology Industry Organization
BNL	—	Brookhaven National Laboratory
BSCS	—	Biological Sciences Curriculum Study
BS/SCF	—	Biological Sequence/Structure Computational Facility
BTCI	—	BioPharmaceutical Technology Center Institute
BTP	—	Biotechnology Training Programs
CAE	—	capillary array electrophoresis
CaSSS	—	California Separation Science Society
CATCMB	—	Center for Advanced Training in Cell and Molecular Biology
CCM	—	Chromosome Coordinating Meeting
CDC	—	Centers for Disease Control
cDNA	—	Complementary DNA
CE	—	Capillary Electrophoresis
CEPH	—	Centre d'Etude du Polymorphisme Humain
CF	—	Cystic Fibrosis
CFF	—	Cystic Fibrosis Foundation
CHH	—	Cartilage-hair Hypoplasia

CHI	—	Cambridge Healthtech Inst.
CHOP	—	Children's Hospital of Philadelphia
CIMB	—	Center for International Meeting on Biology
CIOMS	—	Council for International Organizations of Medical Sciences
CIRB	—	Colorado Institute for Research in Biotechnology
CLMA	—	Clinical Laboratory Management Association
cM	—	Centimorgan
CMT	—	Charcot-Marie-Tooth
CONTIG	—	Consortium of Teachers in Genetics
CORN	—	Council of Regional Networks for Genetic Services
CRADA	—	Cooperative Research and Development Agreement
CSHL	—	Cold Spring Harbor Laboratory
CTG	—	Trinucleotide Repeat
DHHS	—	Dept. of Health and Human Services
DIMACS	—	Center for Discrete Math. & Theoretical Comp. Sci.
DM	—	Myotonic Dystrophy
DMD	—	Duchenne Muscular Dystrophy
DNA	—	Deoxyribonucleic acid
DOE	—	Department of Energy
ECB	—	European Congress on Biotechnology
EDS	—	Electronic Data Submission
EEOC	—	Equal Employment Opportunity Commission

EL.B.A.	—	ELectronics and Biotechnology Advanced
EMBL	—	European Mol. Biol. Lab.
EMBO	—	European Molecular Biology Organisation
EMG	—	Encyclopedia of the Mouse Genome
EMS	—	Environmental Mutagen Society
EORTC	—	European Organization for Research and Treatment of Cancer
ERDA	—	Energy Research and Development Administration
ERI	—	Eleanor Roosevelt Institute
ES	—	Embryonic Stem
ESF	—	European Science Foundation
ESHG	—	European Society of Human Genetics
ESI	—	Electrospray Ionization
EST	—	Expressed Sequence Tag
EUCIB	—	European Collaborative Interspecific Backcross
EURESCO	—	European Research Conferences
FAP	—	Familial Adenomatous Polyposis
FASEB	—	Federation of American Societies for Experimental Biology
FCM	—	Flow Cytometry
FEBS	—	Fed. of Eur. Biochem. Soc.
FMF	—	Familial Mediterranean Fever
FRAXA	—	Fragile X locus
FSHD	—	Facioscapulohumeral Muscular Dystrophy
FTICR	—	Fourier Transform Ioncyclotron Resonance

FVEA	—	Fundacion Valenciana de Estudios Avanzados
GAS	—	Genome Automation System
GBASE	—	Genome Database of the Mouse
GBR	—	Global Business Research
GDB	—	Genome Database
GDB/OMIM	—	Genome Database/Online Mendelian Inheritance in Man
GESTEC	—	Genome Science and Technology Center
GIST	—	Genome Informatics System of Transputers
GLaRGG	—	Great Lakes Regional Genetics Group
GMCRF	—	General Motors Cancer Res. Foundation
GMD	—	Genomic Map Design
GPI	—	Genetics and Public Issues Program
GRAIL	—	Gene Recognition and Analysis Internet Link
GRC	—	Gordon Res. Conf.
GSA	—	Genetics Societies of America
GSDB	—	Genome Sequence Data Base
HAEC	—	Human Artificial Episomal Chromosome
HELSRD	—	Health Effects and Life Science Research Division
HERAC	—	Health and Environmental Research Advisory Committee
HGCC	—	Human Genome Coordinating Committee
HGM	—	Human Genome Meeting

HGMIS	—	Human Genome Management Information System
HGP	—	Human Genome Project
HHMI	—	Howard Hughes Medical Institute
HICSS	—	Hawaii Intl. Conf. on Systems Sci.
HLA	—	Human Leukocyte Antigen
HMDP	—	Homology Database
HNPCC	—	Hereditary Nonpolyposis Colorectal Cancer
HSA	—	Human Serum Albumin
HUGO	—	Human Genome Organisation
IARC	—	International Agency for Research on Cancer
IBC	—	International Business Communications
IBEX	—	International Biotechnology EXpo
IBF	—	International Business Forum
IBI	—	Institute for Biotechnology Information
ICES	—	International Council of Electrophoresis Societies
ICGEB	—	International Centre for Genetic Engineering and Biotechnology
ICHG	—	International Congress of Human Genetics
ICPEMC	—	International Commission on Protection Against Environmental Mutagens and Carcinogens
ICRF	—	Imperial Cancer Research Fund
IEEE	—	Institute of Electrical and Electronics Engineers
IG	—	IntelliGenetics

IGES	—	International Genetics Epidemiology Societies
IJCAI	—	International Joint Conference on Artificial Intelligence
IMA	—	Institute for Mathematics and its Applications
IMACS	—	International Association for Mathematics and Computers In Simulation
IMAGE	—	Integrated Molecular Analysis of Gene Expression
IMGS	—	Intl. Mammalian Genome Society
INRIA	—	French Natl. Inst. for Research in Computer Science and Control
IOM	—	Institute of Medicine
IOR	—	Institute of Religion
ISAG	—	Intl. Society for Animal Genetics
ISMB	—	Intelligent Systems for Molecular Biology
ISONG	—	Intl. Soc. of Nurses in Genet.
ISQL	—	Interactive Standard Query Language
ISTR	—	Institute for Science Training and Research
IUBMB	—	Intl. Union of Biochemistry and Molecular Biology
IU	—	Indiana University
IVD	—	*in vitro* Diagnostic
JGI	—	Joint Genome Institute
LANL	—	Los Alamos National Laboratory
LBNL	—	Lawrence Berkeley National Laboratory

LLNL	—	Lawrence Livermore National Laboratory
LTI	—	Life Technologies, Inc.
MALDI	—	matrix-assisted laser desorption ionization
MBC	—	Massachusetts Biotechnology Council
MBL	—	Marine Biological Laboratory
MCD	—	Mouse Cytogenetic Database
MDA	—	Muscular Dystrophy Assoc.
MEMS	—	Microelectromechanical Systems
MGC	—	Mouse Genome Conference
MGD	—	Mouse Genome Database
MGI	—	Microbial Genome Initiative
MHC	—	Major Histocompatibility Complex
MIMBD	—	Meet. on the Interconnection of Mol. Biol. Databases
MIT	—	Massachusetts Institute of Technology
MMRF	—	Marshfield Medical Research Foundation
MNBWS	—	Miami Nature Biotechnology Winter Symposium
MOD	—	March of Dimes
MOU	—	Memorandum of Understanding
MRC	—	Medical Research Council
MS	—	Mass Spectrometry
NACHGR	—	Natl. Advisory Council for Human Genome Res.
NAPBC	—	Natl. Action Plan on Breast Cancer
NAS	—	Natl. Academy of Sciences

NASA	—	Natl. Aeronautics and Space Administration
NBAC	—	Natl. Bioethics Advisory Commission
NCGR	—	National Center for Genome Resources
NCI	—	Natl. Cancer Institute
NCSA	—	National Center for Supercomputing Applications
NCSL	—	National Conference of State Legislatures
NCSU	—	North Carolina State University
NEH	—	National Endowment for the Humanities
NFCR	—	National Foundation for Cancer Research
NFID	—	National Foundation for Infectious Diseases
NHGRI	—	Natl. Human Genome Res. Inst.
NICHD	—	National Institute of Child Health and Human Development
NIGMS	—	National Institute of General Medical Sciences
NIH	—	National Institutes of Health
NIGMS	—	Natl. Institute of General Medical Sciences
NIST	—	Natl. Inst. of Standards and Technology
NLGLP	—	National Laboratory Gene Library Project
NLM	—	National Library of Medicine

NMHCC	—	National Managed Health Care Congress
NORD	—	National Organization for Rare Disorders
NRC	—	National Research Council
NRSA	—	National Research Service Award
NSF	—	Natl. Sci. Foundation
NSGC	—	National Society of Genetic Counselors
NSTA	—	National Science Teachers Association
NYAS	—	New York Academy of Science
OAINN	—	Ohio Aerospace Institute Neural Networks
OBER	—	Office of Biological and Environmental Research
OHER	—	Office of Health and Environmental Research
OMIM	—	Online *Mendelian Inheritance in Man*
OPRR	—	Office of Protection from Res. Risks
ORAU	—	Oak Ridge Associated Universities
ORISE	—	Oak Ridge Inst. for Science and Education
ORF	—	open reading frame
ORNL	—	Oak Ridge National Laboratory
OSU	—	Oregon State University
OTA	—	Office of Technology Assessment
PACHG	—	Program Advisory Committee on the Human Genome
PCR	—	Polymerase Chain Reaction

PFGE — Pulsed-field gel electrophresis
PG — Plant Genome
PIR — Protein Info. Resource
PNNL — Pacific Northwest National Laboratory
PRIM&R — Public Responsibility in Medicine & Research
PSC — Pittsburgh Supercomputing Center
QDFM — Quantitative DNA Fiber Mapping
RAPD — Random Amplified Polymorphic DNA
RARA — Retinoic Acid Receptor
RECOMB — Conference on Computational Molecular Biology
RFLP — Restriction Fragment Length Polymorphism
RH — Radiation Hybrid
RLGS — Restriction Landmark Genomic Scanning
SASE — Sample Sequencing
SBH — Sequencing by Hybridization
SBIR — Small Business Innovation Research
SCAN — Sequence Comparison ANalysis Program
SCE — School of Continuing Education
SCI — Society of Chemical Industry
SCW — Single-chromosome Workshop
SDC — San Diego Conference
SERGG — Southeast Regional Genetics Group
SHOM — Sequencing by Hybridization on Matrices

SIAM	—	Society for Industrial and Applied Mathematics
SIB	—	Society for Industrial Biology
SIM	—	Society for Industrial Microbiology
SIMS	—	Societal Institute of the Mathematical Sciences
SIVB	—	Society for In Vitro Biology
SNL	—	Sandia National Laboratory
SNP	—	Single Nucleotide Polymorphism
snRNP	—	Small Nuclear Ribonucleo Protein Particle
SQL	—	Standard Query Language
SSLP	—	Single-sequence Length Polymorphism
SSR	—	Simple Sequence Repeat
STM	—	Scanning Tunneling Microscope
STRP	—	Short Tandem Repeat Polymorphism
STS	—	Sequence-tagged Site
SULS	—	Suffolk Univ. Law School
TCR	—	T-cell receptor
TIGR/NIST	—	The Inst. for Genomic Res./Natl. Inst. of Standards and Technol.
UC	—	University of California
UCB	—	University of California, Berkeley
UCLA	—	University of California, Los Angeles
UICM	—	Univ. of Illinois College of Medicine
UMBC	—	University of Maryland, Baltimore County
UNC	—	University of North Carolina at Chapel Hill

UMDS	—	United Medical & Dental Schools (Univ. of London)
UNESCO	—	United Nations Educational, Scientific, and Cultural Organization
USDA	—	U.S. Dept. of Agriculture
UT	—	University of Tennessee
UW	—	University of Washington
WCCS	—	Workshop on Complete cDNA Sequencing
WHS	—	Wolf-Hirschhorn syndrome
WLMG	—	Wellcome Laboratory for Molecular Genetics
WSES	—	World Scientific and Engineering Society

List of Glossary of Terms

Alternative mRNA splicing: The inclusion or exclusion of different exons to form different mRNA transcripts.

Amino acid: Any of 20 basic building blocks of proteins—composed of a free amino (NH2) end, a free carboxyl (COOH) end, and a side group (R).

Ampicillin (beta-lactamase): An antibiotic derived from penicillin that prevents bacterial growth by interfering with cell wall synthesis.

Amplify: To increase the number of copies of a DNA sequence, in vivo by inserting into a cloning vector that replicates within a host cell, or in vitro by polymerase chain reaction.

Anaerobe: An organism that grows in the absence of oxygen.

Analyte: A chemical species targeted for qualitative or quantitative analysis.

Anneal: The pairing of complementary DNA or RNA sequences, via hydrogen bonding, to form a double-stranded polynucleotide. Most often used to describe the binding of a short primer or probe.

Antibiotic resistance: The ability of a microorganism to produce a protein that disables an antibiotic or prevents transport of the antibiotic into the cell.

Antibiotic: A class of natural and synthetic compounds that inhibit the growth of or kill other microorganisms.

Antibody: An immunoglobulin protein produced by Blymphocytes of the immune system that binds to a specific antigen molecule.

Anticodon: A nucleotide base triplet in a transfer RNA molecule that pairs with a complementary base triplet, or codon, in a messenger RNA molecule.

Antigen: Any foreign substance, such as a virus, bacterium, or protein, that elicits an immune response by stimulating the production of antibodies.

Antigenic determinant: A surface feature of a microorganism or macromolecule, such as a glycoprotein, that elicits an immune response.

Antigenic switching: The altering of a microorganism's surface antigens through genetic rearrangement, to elude detection by the host's immune system.

Antimicrobial agent: Any chemical or biological agent that harms the growth of microorganisms.

Antisense DNA: DNA strand that is complementary (opposite) to the functional gene. Antisense DNA can block the function of the normal sense DNA. This method was used in the construction of the Flavr Savr tomato.

Bacteriophage (phage): Any virus that infects bacteria. They were the first organisms used for the study of molecular genetics and are now widely used as cloning vectors.

Bacteriorhodopsin: Pigmented protein found in the plasma membrane of a salt-loving bacterium, *Halobacterium halobium*; it pumps protons out of the cell in response to light.

Bacterium (plural bacteria): Common name for any member of the diverse group of procaryotic organisms. Most are single cells, but multicellular forms also exist.

Base pair (bp): A pair of complementary nitrogenous bases in a DNA molecule—adenine-thymine and guanine-cytosine. Also, the unit of measurement for DNA sequences. In DNA, there are four possible bases: cytosine (C), guanine (G), adenine (A), and thymine (T). Cytosine and thymine are pyrimidine bases; adenine and guanine are purine bases. Cytosine is complementary to guanine while adenine is complementary to thymine. If one strand of DNA has the sequence ATTGC then the complementary strand will be TAACG. Two complementary bases constitute a base pair. In RNA, thymine is replaced by uracil.

Beta-DNA: The normal form of DNA found in biological systems, which exists as a right-handed helix.

Binding energy: The reduction in the free energy of a system that occurs when a ligand binds to a receptor. Generally used to describe the total energy required to remove something, or to take a system apart into its constituent particles—for example, to separate two atoms from one another, or to separate an atom into electrons and nuclei.

Bio-assemblies or Biomolecular Assemblies: Containing several protein units, DNA loops, lipids, various ligands, etc.

Bioaugmentation: Increasing the activity of bacteria that decompose pollutants; a technique used in bioremediation.

Biochauvinism: The prejudice that biological systems have an intrinsic superiority that will always give them a monopoly on self-reproduction and intelligence.

Biocompatible coated materials: Biocompatible materials usually used in dental and bone implants that enhance biologic fixation, thereby increasing the bond strength between the coated material and bone, and minimize possible biological effects that may result from the implant itself. MeSH, 1999

Biocompatible materials: Synthetic or natural materials, other than drugs, that are used to replace or repair any body tissue or bodily function. MeSH, 1973

Biodefense: The need for improved national defenses against biological attacks are urgently needed. A real-time reporting system needs to be developed, whereby officials can be informed about an emerging threat before an agent has a chance to affect thousands of people. The development of integrated systems for detecting and monitoring biological agents is instrumental to this goal. The development of vaccines and novel detection platforms will enable preparedness, while also benefiting human health.

Biodiversity: The wide diversity and interrelatedness of earth organisms based on genetic and environmental factors.

Biodynotics Biologically Inspired Multifunctional Dynamic Robotics: A multidisciplinary, multi-pronged approach with far reaching impact on robotic capabilities for national security applications. Biologically Inspired Multifunctional Dynamic Robotics: BIODYNOTICS will explore the following areas: DYNAMIC MOBILITY, BEHAVIOR, INTEGRATION Biological Sciences, Defense Sciences Office, DARPA.

Bioengineering: Is rooted in physics, mathematics, chemistry, biology, and the life sciences. It is the application of a systematic, quantitative, and integrative way of thinking about and approaching the solutions of problems important to biology, medical research, clinical proactive, and population studies. The NIH Bioengineering Consortium agreed on the following definition for bioengineering research on biology, medicine, behavior, or health recognizing that no definition could completely eliminate overlap with other research disciplines or preclude variations in interpretation by different individuals and organizations. Integrates physical, chemical, or mathematical sciences and engineering principles for the study of biology, medicine, behavior, or health. It advances fundamental concepts, creates knowledge for the molecular to the organ systems levels, and develops innovative biologics, materials, processes, implants, devices, and informatics approaches for the prevention, diagnosis, and treatment of disease, for patient rehabilitation, and for improving health.

Bioenrichment: Adding nutrients or oxygen to increase microbial breakdown of pollutants.

Biofabrication: Includes nanoparticle delivery systems, biomaterials, tissue engineering, implants and prostheses. Using biological processes to synthesize and manufacture chemicals and materials of high value to the Department of Defense. Biological processes are characterized by: low energy barriers (~10Kcal, high temperatures/pressures not required); high reaction-, regio- and stereo- specificity (protection/ deprotection wastes and catalyst poisoning avoided); spatio-temporal control of materials synthesis of defined composition and size with angstrom- level precision; and local control of the dielectric environment (largely eliminating the need for toxic solvents). Potential target materials and chemicals include composites with enhanced mechanical properties, ultra-low k dielectrics and thermoelectrics, optoelectronic materials, photonic devices (waveguides and logic elements), electronic materials (e.g., GaN, InGaN, AlGaN), elastomers, energetic materials, and adhesives. Biofabrication, DARPA.

Biologics: Agents, such as vaccines, that give immunity to diseases or harmful biotic stresses.

Biomass: The total dry weight of all organisms in a particular sample, population, or area.

Biomaterials: Synthetic or natural materials that can replace or augment tissues, organs or body functions.

Biomechanics: Mechanical structures of living organisms (especially muscles and bones).

Biomedical Nanotechnology: (1) the comprehensive monitoring, control, construction, repair, defense, and improvement of all human biological systems, working from the molecular level, using engineered nanodevices and nanostructures; (2) the science and technology of diagnosing, treating, and preventing disease and traumatic injury, of relieving pain, and of preserving and improving human health, using molecular tools and molecular knowledge of the human body; (3) the employment of molecular machine systems to address medical problems, using molecular knowledge to maintain and improve human health at the molecular scale. (4) A rapidly expanding field that includes many potential technologies and approaches. The key to this definition is that phenomena and materials at the nanometer scale are known to have properties that are uniquely attributable to that scale length. Nanomedicine could similarly be defined [as nanotechnology] as the design, synthesis, or application of materials, devices, or technologies in the nanometer-scale for the basic understanding, diagnosis, and / or treatment of disease. Canadian Institute of Health Research, Regenerative Medicine and Nanomedicine, RFA, 2003. The monitoring, repair, construction and control of human biological systems at the molecular level, using engineered nanodevices and nanostructures. (5) the comprehensive monitoring, control, construction, repair, defense, and improvement of all human biological systems, working from the molecular level, using engineered nanodevices and nanostructures; (6) the science and technology of diagnosing, treating, and preventing disease and traumatic injury, of relieving pain, and of preserving and improving human health, using molecular tools and molecular knowledge of the human body; (7) the employment of molecular machine systems to address medical problems, using molecular knowledge to maintain and improve human health at the molecular scale. The use of molecular-scale devices (nanotechnology) to repair damage and boost the immune system. Vapour grown carbon fibres are obtained as shown in this SEM image. The diameters of these fibres can vary from 100 nm to 500 nm. As part of the National Institutes of Health (NIH) Roadmap for Medical Research [nihroadmap.nih. gov], the NIH [nih.gov] will establish a handful of nanomedicine centers. These centers will be staffed by a highly interdisciplinary scientific crew including biologists, physicians, mathematicians, engineers and computer scientists. Research conducted over the first few years will be spent gathering extensive information about how molecular machines are built. A key activity during this time will be the development of a new kind of vocabulary, or lexicon, to define biological parts and processes in engineering terms. Once researchers have completely catalogued the interactions between and within molecules, they can begin to look for patterns and a higher order of connectedness than is possible to identify with current experimental methods. Mapping these networks and understanding how they change over time will be a crucial step toward helping scientists understand nature's rules of biological design. Understanding these rules will, in many years' time, enable researchers to use this information to address biological issues in unhealthy cells.

Biomedical polymers: More and more therapeutic problems are relevant to the use of polymer- based therapeutic aids for a limited period of time, namely the healing time related to the outstanding capacity of living systems to self- repair ... After healing the remaining prosthetic materials or devices become foreign residues or wastes that have to be eliminated from the body. Nowadays, biocompatible polymers that can degrade in the body are developed. The degradation and the elimination of degradation by-products depend on rather complex phenomena that are presently reflected inconsistently by terms issued from the tradition because each domain has developed its own terminology almost independently. This is a source of misunderstandings, confusions and misperceptions among scientists, surgeons, pharmacists, journalists and politicians, the situation being increased by the introduction of degradable polymers in plastic waste management and environmental protection. Therefore, it is urgent to reflect the various phenomena by specific terms, harmonize and enforce their use by the people active in the biomedical, pharmacological and environmental fields, and, last but not least by the publishing media and journalists. IUPAC, Terminology for biomedical (therapeutic) polymers.

BioMEMS Biological MicroElectro Mechanical Systems: Highlights the technical advances in the field that are leading a revolution in medicine, and creating a new generation of analytical devices for medical diagnostics. The meeting will encompass technology developments in micro & nano drug delivery, interface of nanotech and tissue engineering, microfluidics, and miniaturized total analysis systems (microTAS), biosensors, innovations in mass spec, and nanoscale imaging *BioMEMS and Nanotech World*, Aug. 16- 17, 2004, Washington DC

BioMEMS: MEMS used in medicine, that use microchips.

Biomimetic Chemistry: Knowledge of biochemistry, analytical chemistry, polymer science, and biomimetic chemistry is linked and applied to research in designing new molecules, molecular assemblies, and macromolecules having biomimetic functions. These new bio-related materials of high performance, including, for example, enzyme models, synthetic cell membranes, and biodegradable polymers, are prepared, tested, and constantly improved in this division for industrial scale production.

Biomimetic materials: Materials fabricated by BIOMIMETICS techniques, i.e., based on natural processes found in biological systems. [MeSH 2003] (2) Materials that imitate, copy, or learn from nature.

Biomimetic: Imitating, copying, or learning from nature. Nanotechnology already exists in nature; thus, nanoscientists have a wide variety of components and tricks already available. An interdisciplinary field in materials science, ENGINEERING, and BIOLOGY, studying the use of biological principles for synthesis or fabrication of BIOMIMETIC MATERIALS. There is a need to develop the next generation of restorative materials and medical implants. New avenues of scientific inquiry may enable the development of biomaterials that are safe, reliable, "smart", long-lasting, and perform ideally in their respective biological environments.... Over the last few years biomimetics and tissue engineering have emerged as a new vision in the field of tissue and organ repair and restoration. Biomimetics and tissue engineering are interdisciplinary fields that combine information from the study of biological structures and their functions with physics, mathematics, chemistry and engineering for the generation of new materials, tissues and organs. In the area of craniofacial, oral and dental principles from biomimetics and tissue engineering are applied to developing dental and facial implants, new polymers for guided tissue regeneration used in treating periodontal disease and bone and connective tissue defects, coral- based hydroxyapatite replicas for reconstruction of alveolar ridges and other osseous defects, temporomandibular joint (TMJ) and other joint prostheses, formation of bone matrix substitutes, and artificial replicas of bone, skin, and mucosa. The term biomimetics was coined in 1972 in the context of artificial enzymes; it might be defined broadly as "the abstraction of good design from nature". It is a fact of everyday life that nature has managed to built materials and 'devices' with breathtaking functionality, heterogeneity and stability by using a comparatively limited number of building blocks (the whole range of synthetic materials is restricted to man- made engineering). The basic concepts of nature are often simple; it is the way in which building blocks and materials are arranged that results in functionality. Among the most simple and abundant themes of nature is self- assembly: lipids assemble in sheets to form cell membranes, proteins assemble into functional enzymes, cellular 'sensors', fibers, or virus coats, and DNA assembles in double strands to provide the very basis for live: replication. Biomimetics Study of the structure and function of biological substances to make artificial products that mimic the natural ones.

BIOML Biopolymer Markup Language: Designed by the BIOML core team at Proteometrics, LLC and Proteometrics Canada Ltd. It is to be used "for the annotation of biopolymer sequence information. BIOML allows the full specification of all experimental information known about molecular entities composed of biopolymers, for example, proteins and genes."

Biomolecular materials: An emerging discipline, materials whose properties are abstracted from biology. They share many of the characteristics of biological materials but are not necessarily of biological origin. For example, they may be inorganic materials that are organized or processed in a biomimetic fashion. A key feature of biological and biomolecular materials is their ability to undergo *self- assembly*.

Biomolecular Nanotechnology: Nanotechnology existing in living systems and resulting from our ability to use biomolecules as components for molecular nanotechnology.

Biomotors: Driven by energy sources such as *adenosine triphosphate (ATP)* for chemical transduction and other processes. These biomotors are considered to be biomolecular and are discussed in the body of this report, but strictly speaking they do not conform to the panel's definition of self- assembly.

Bionanotechnology: Includes molecular motors, biomaterials, single molecule manipulation technologies, biochip technologies, etc.

Biopolymeroptoelectromechanical Systems [BioPOEMS]: Combining optics and microelectromechanical systems, and used in biological applications.

Biopolymers: Macromolecules (including proteins, nucleic acids and polysaccharides) formed by living organisms.

Bioremediation: The use of microorganisms to remedy environmental problems. See Bioaugmentation, Bioenrichment.

Biorobotics: Our research focuses on the role of sensing and mechanical design in motor control, in both robots and humans. This work draws upon diverse disciplines, including biomechanics, systems analysis, and neurophysiology. The main approach is experimental, although analysis and simulation play important parts. In conjunction with industrial partners, we are developing applications of this research in biomedical instrumentation, teleoperated robots, and intelligent sensors.

Biosensor: The term "biosensor" is a general designation that denotes either a sensor to detect a biological substance or a sensor which incorporates the use of biological molecules such as antibodies or enzymes. Biosensors are a subcategory of chemical sensors.

Biostasis: A condition in which an organism's cell and tissue structure are preserved, allowing later restoration by cell repair machines. Applicable to cryonics.

Biotechnology: The scientific manipulation of living organisms, especially at the molecular genetic level, to produce useful products. Gene splicing and use of recombinant DNA (rDNA) are major techniques used. Technology for working with biological systems. Includes genetic engineering, human and veterinary medicine, crop and animal breeding, diagnostics, pharmaceuticals, forensics, etc. Narrow sense: Genetic engineering.

Biotic stress: Living organisms which can harm plants , such as viruses, fungi, and bacteria, and harmful insects.

Biovorous: From "biovore;" an organism capable of converting biological material into energy for sustenance.

Bipolar-junction transistor: Transistor with n-type and p-type semiconductors having base-emitter and collector-base junctions.

Blocks Substitution Matrix: A substitution matrix in which scores for each position are derived from *observations* of the frequencies of substitutions in blocks of local alignments in related proteins. Each matrix is tailored to a particular evolutionary distance. In the BLOSUM62 matrix, for example, the alignment from which scores were derived was created using sequences sharing no more than 62% identity. Sequences more identical than 62% are represented by a single sequence in the alignment so as to avoid over-weighting closely related family members. (Henikoff and Henikoff)

Blot, northern: A pattern of RNA fragments transferred to a nitrocellulose membrane from a gel. The gel has undergone electrophoresis to separate fragments according to size. The RNA fragments are arranged in different lanes for each sample. Each lane contains bands which are fragments of different sizes. Radioactive probes are often used to visualized particular bands by autoradiography.

Blot, southern: A pattern of DNA fragments transferred to a nitrocellulose membrane from a gel. The gel has undergone electrophoresis to separate fragments according to size. The DNA fragments are arranged in different lanes for each sample. Each lane contains bands which are fragments of different sizes. Radioactive probes are often used to visualized particular bands by autoradiography.

Blot, western: A pattern of proteins transferred to a

nitrocellulose membrane from a gel. The gel has undergone electrophoresis to separate fragments according to size. The proteins are arranged in different lanes for each sample. Each lane contains bands which are fragments of different sizes. Labeled antibody probes are often used to visualize particular bands.

Bone substitutes: Synthetic or natural materials for the replacement of bones or bone tissue. They include hard tissue replacement polymers, natural coral, hydroxyapatite, beta- tricalcium phosphate, and various other biomaterials. The bone substitutes as inert materials can be incorporated into surrounding tissue or gradually replaced by original tissue.

Brownian Assembly: Brownian motion in a fluid brings molecules together in various position and orientations. If molecules have suitable complementary surfaces, they can bind, assembling to form a specific structure. Brownian assembly is a less paradoxical name for self-assembly (how can a structure assemble itself, or do anything, when it does not yet exist?).

Bt (*Bacillus thuringiensis*): A soil bacterium that produces insecticidal proteins. There are several different kinds of proteins produced by different strains of Bt. Some are effective against larvae of moths and butterflies. Others are effective against larvae of beetles. The Bt protein has been introduced into various crops as a built-in insecticide.

Calorie: Unit of heat. One calorie (small "c") is the amount of heat needed to raise the temperature of 1 gram of water by 1 °C. A kilocalorie is the unit used to describe the energy content of foods.

CAP (catabolite gene activator protein): Gene regulatory protein in procaryotes that, when glucose is absent, activates genes responsible for the breakdown of alternative carbon sources.

Capacitor: Energy storage circuit element having two conductors separated by an insulator.

Carcinogen: A substance that induces cancer.

Carcinoma: A malignant tumor derived from epithelial tissue, which forms the skin and outer cell layers of internal organs.

Catalyst: A substance that promotes a chemical reaction by lowering the activation energy of a chemical reaction, but which itself remains unaltered at the end of the reaction.

Catalytic antibody (abzyme): An antibody selected for its ability to catalyze a chemical reaction by binding to and stabilizing the transition state intermediate.

Catalytic RNA (ribozyme): A natural or synthetic RNA molecule that cuts an RNA substrate.

cDNA library: A library composed of complementary copies of cellular mRNAs.

cDNA: DNA synthesized from an RNA template using reverse transcriptase. Complementary DNA to a particular RNA fragment.

Cell pharmacology: Delivery of drugs by medical nanomachines to exact locations in the body.

Cell Repair Machine: A system including nanocomputers and molecular scale sensors and tools, programmed to repair damage to cells and tissues. Molecular and nanoscale machines with sensors, nanocomputers and tools, programmed to detect and repair damage to cells and tissues, which could even report back to and receive instructions from a human doctor if needed.

Cell surgery: In medical nanorobotics, modifying cellular structures using medical nanomachines. Modifying cellular structures using medical nanomachines.

Cell: A membrane-bound unit, typically microns in diameter. All plants and animals are made up of one or more cells (trillions, in the case of human beings). In general, each cell of a multicellular organism contains a nucleus holding all of the genetic information of the organism. (2) A small structural unit, surrounded by a membrane, making up living things.

Cellular oncogene (proto-oncogene): A normal gene that when mutated or improperly expressed contributes to the development of cancer.

Centers of origin: Usually the location in the world where the oldest cultivation of a particular crop has been identified.

Central dogma: Francis Crick's seminal concept that in nature genetic information generally flows from DNA to RNA to protein.

Centrifugation: Separating molecules by size or density using centrifugal forces generated by a spinning rotor. G forces of several hundred thousand times gravity are generated in ultracentrifugation.

Centromere: The central portion of the chromosome to which the spindle fibers attach during mitotic and meiotic division.

Chemotherapy: A treatment for cancers that involves administering chemicals toxic to malignant cells.

Chloramphenicol: An antibiotic that interferes with protein synthesis.

Chromatid: Each of the two daughter strands of a duplicated chromosome joined at the centromere during mitosis and meiosis.

Chromosome walking: Working from a flanking DNA marker, overlapping clones are successively identified that span a chromosomal region of interest.

Chromosome: A single DNA molecule, a tightly coiled strant of DNA, condensed into a compact structure in vivo by complexing with accessory histones or histone-like proteins. Chromosomes exist in pairs in higher eukaryotes. A linear structure in the nucleus of plants and animals that is visible in light microscopy when stained. The chromosome is a single, long, linear molecule of DNA and associated proteins. Bacteria have a single circular chromosome; other organisms may have many linear chromosomes.

Cistron: A DNA sequence that codes for a specific polypeptide; a gene. See DNA, Gene.

Clone: An exact genetic replica of a specific gene or an entire organism. See Cloning.

Cloning vector: A DNA molecule capable of autonomous replication within the cloning host cell (e.g *E. coli*). The vectors contain restriction enzyme sites for insertion of foreign DNA. Cloning vectors are derived from bacterial plasmids, bacteriophages, or viruses.

Cloning: The mitotic division of a progenitor cell to give rise to a population of identical daughter cells or clones.

Coat protein (capsid): The coating of a protein that enclosed the nucleic acid core of a virus.

Codon: A group of three nucleotides that specifies addition of one of the 20 amino acids during translation of an mRNA into a polypeptide. Strings of codons form genes and strings of genes form chromosomes. Set of three nucleotides that specify a particular amino acid during protein synthesis.

Coenzyme (cofactor): An organic molecule, such as a vitamin, that binds to an enzyme and is required for its catalytic activity.

Colony: A group of identical cells (clones) derived from a single progenitor cell.

Combinatorial materials design: Uses computing power (sometimes together with massive parallel experimentation) to screen many different materials possibilities to optimize properties for specific applications (e.g., catalysts, drugs, optical materials).

Commensalism: The close association of two or more dissimilar organisms where the association is advantageous to one and doesn't affect the other(s). See Parasitism, Symbiosis.

Competency: An ephemeral state, induced by treatment with cold cations, during which bacterial cells are capable of uptaking foreign DNA.

Complementary DNA or RNA: The matching strand of a DNA or RNA molecule to which its bases pair.

Complementary nucleotides: Members of the pairs adenine-thymine, adenine-uracil, and guaninecytosine that have the ability to hydrogen bond to one another.

Concatemer: A DNA segment composed of repeated sequences linked end to end.

Configuration space: A mathematical space describing the three-dimensional configuration of a system of particles (e.g., atoms in a nanomechanical structure) as a single point; the configuration space for an N particle system has $3N$ dimensions.

Conformation: The shape of a molecule. A molecular geometry that differs from other geometries chiefly by rotation about single or triple bonds; distinct conformations (termed con formers) are associated with distinct potential wells. Typical biomolecules and products of organic synthesis can interconvert among many conformations; typical diamondoid structures are locked into a single potential well, and thus lack conformational flexibility.

Conjugation: The joining of two bacteria cells when genetic material is transferred from one bacterium to another.

Constitutive promoter: An unregulated promoter that allows for continual transcription of its associated gene.

Contiguous (contig) map: The alignment of sequence data from large, adjacent regions of the genome to produce a continuous nucleotide sequence across a chromosomal region.

Coulomb [C]: Measure of electrical charge: 1 C is an amount of charge equal to that of about 6.24x1018 electrons.

Coupling: A connection between more than one object or energy pathway so that together they function as a single unit.

Cross-hybridization: The hydrogen bonding of a single-stranded DNA sequence that is partially but not entirely complementary to a singlestranded substrate. Often, this involves hybridizing a DNA probe for a specific DNA sequence to the homologous sequences of different species.

Crossing-over: The exchange of DNA sequences between chromatids of homologous chromosomes during meiosis.

Cross-pollination: Fertilization of a plant from a plant with a different genetic makeup.

Crosstalk: Electromagnetic noise transmitted between leads or circuits in close proximity to each other.

Crybiology: The science of biology at low temperatures; research in cryobiology has made possible the freezing and storing of sperm and blood for later use.

Crystal Lattice: The regular three-dimensional pattern of atoms in a crystal.

Crystal structure: For crystalline materials, the manner in which atoms or ions are arrayed in space. It is defined in terms of the unit cell geometry and the atom positions within the unit cell.

Crystallescence: In medical nanorobotics, the crystallization of solid solute that is offloaded by nanorobot sorting rotors at a concentration that exceeds the solvation capacity of the surrounding solvent.

Culture: A particular kind of organism growing in a laboratory medium.

Curie temperature (also Curie point) (Tc): The temperature above which a ferromagnetic or ferrimagnetic material becomes paramagnetic. For iron the Curie point is 760oC and for nickel 356oC.

Current [A]: Measure of rate of flow of electric charge: a one-ampere current is a flow of 1 C of charge per second.

Cutoff: Condition in a diode or bipolar junction transistor in which the potential across a p-n junction prevents current flow.

Cyclic AMP (cyclic adenosine monophosphate): A second messenger that regulates many intracellular reactions by transducing signals from extracellular growth factors to cellular metabolic pathways.

Cyclic: A structure is termed cyclic if its covalent bonds form one or more rings.

Cytogenetics: Study that relates the appearance and

behavior of chromosomes to genetic phenomenon.

Dalton: A unit of measurement equal to the mass of a hydrogen atom, 1.67 x 10E-24 gram/L (Avogadro's number).

Denature: To induce structural alterations that disrupt the biological activity of a molecule. Often refers to breaking hydrogen bonds between base pairs in double-stranded nucleic acid molecules to produce in single-stranded polynucleotides or altering the secondary and tertiary structure of a protein, destroying its activity.

Density gradient centrifugation: High-speed centrifugation in which molecules "float" at a point where their density equals that in a gradient of cesium chloride or sucrose.

Diabetes: A disease associated with the absence or reduced levels of insulin, a hormone essential for the transport of glucose to cells.

Dideoxynucleotide (didN): A deoxynucleotide that lacks a 3' hydroxyl group, and is thus unable to form a 3'-5' phosphodiester bond necessary for chain elongation. Dideoxynucleotides are used in DNA sequencing and the treatment of viral diseases.

Digest: To cut DNA molecules with one or more restriction endonucleases.

Diploid cell: A cell which contains two copies of each chromosome.

Directional cloning: DNA insert and vector molecules are digested with two different restriction enzymes to create noncomplementary sticky ends at either end of each restriction fragment. This allows the insert to be ligated to the vector in a specific orientation and prevents the vector from recircularizing.

DNA: DNA is the recipe for life. DNA is a molecule found in the nucleus of every cell and is made up of 4 nucleotides Adenine (A), Guanine (G), Thymine (T), and Cystosine (C). The order of the nucleotides in the DNA strand holds a code of information for the cell.

DNA (Deoxyribonucleic acid): An organic acid and polymer composed of four nitrogenous bases—adenine, thymine, cytosine, and guanine linked via intervening units of phosphate and the pentose sugar deoxyribose. DNA is the genetic material of most organisms and usually exists as a double-stranded molecule in which two antiparallel strands are held together by hydrogen bonds between adeninethymine and cytosine-guanine.

DNA (Deoxyriobonucleic Acid): DNA molecules are long chains consisting of four kinds of nucleotides; the order of these nucleotides encodes the information needed to construct protein molecules. These in turn make up much of the molecular machinery of the cell. DNA is the genetic material of cells.

DNA Chip: Also: Gene Chip and DNA Microchip. A purpose built microchip used to identify mutations or alterations in a gene's DNA.

DNA diagnosis: The use of DNA polymorphisms to detect the presence of a disease gene.

DNA diagnostics - miniaturization of: In the areas of sample preparation and assay, it is clear that miniaturization is key. To reduce the size of samples by a factor of 10 or greater, barriers in microfluidics, micromachining, robotics, microchemistry, nucleic acid chemistry, and surface chemistry must be overcome. To implement miniaturized protocols accurately and efficiently, substantial automation of the process will be required. In the development of miniaturized systems, it is essential that the system can be adapted for high levels of parallelization. Miniaturization poses significant technological risks. Currently, there exists no universally accepted precedent for the handling, replication, amplification, or cloning of DNA in nanoliter volumes. Due to the size and charge of the DNA molecule, and the relative instability of many of the enzymes involved in the sample preparation processes, nanoliter and less volumes may pose substantial challenges. In addition, interactions of the biological molecules with the surfaces of the reaction chambers must be minimized. For some methodologies, it is not clear what the optimal sample will be, so substantial improvement in DNA fragmentation technologies or DNA cloning vectors may be required for the ultimate efficient application to diagnostics. Improvements in any of these areas are likely to be of value to other non- DNA based diagnostic applications such as antibody screening protocols and enzyme based diagnostics, because miniaturized robotic or micro- electro mechanical systems developed for DNA could be modified to be used for these purposes.

DNA fingerprint: The unique pattern of DNA fragments identified by Southern hybridization (using a probe that binds to a polymorphic region of DNA) or by polymerase chain reaction (using primers flanking the polymorphic region).

DNA polymorphism: One of two or more alternate forms (alleles) of a chromosomal locus that differ in nucleotide sequence or have variable numbers of repeated nucleotide units.

DNA Probes: A DNA or nucleic acid probe is a short strand of DNA that locates and binds to its complementary sequence in samples containing single strands of DNA or RNA enabling identification of specific sequences. Nucleic acid probe assays exploit the fundamental hybridization reaction that occurs spontaneously between two complementary DNA:DNA or DNA:RNA strands. As in immunoassays, detection of the hybrid requires that the probe be labeled. Various direct and indirect methods have been devised for the detection of the hybrid. Direct labeling involves attaching the label directly to the probe sequence; indirect labeling binds an antibody to the DNA:DNA or DNA:RNA hybrid. As in immunoassays, non-isotopically-labeled probes are preferred over radio-labeled probes primarily because of radiation hazards, disposal problems, and short reagent shelf life. In addition, the factors determining the detection limits of hybridization assays based on labeled probes are similar to those in immunoassays. Therefore, the development of a simple, inexpensive and sensitive direct detection system which eliminates the use of labels is highly desirable.

DNA Sequencing: There are two main classical methods for sequencing DNA: The first method, developed by Allan Maxam and Walter Gilbert, involves chemicals used to cleave the DNA at certain positions, generating a set of fragments that differ by one nucleotide. The second method, developed by Fred Sanger and Alan Coulson, involves enzymatic synthesis of DNA strands that terminate in a modified nucleotide. Analysis of fragments is similar for both methods and involves gel electrophoresis and autoradiography or fluorescence. The enzymatic method has largely replaced the chemical method as the technique of choice, although there are some situations where chemical sequencing can provide data more easily than the enzymatic method.

DNA, recombinant: DNA that has been cut and spliced back together in a new sequence. The DNA may be from one organism or from more than one organism

DNA, repetitive: Fragments of DNA that appear in multiple copies in a single individual.

DNA: A molecule encoding genetic information, found in the cell's nucleus. Deoxyribose Nucleic Acid. DNA is a molecule used in biological systems to store genetic information. It has a long, double stranded structure with the two strands wrapped around each other to form a right handed helix. There are about 10 nucleotide pairs per helical turn. Each strand is composed of a sugar phosphate backbone and attached bases. It is connected to a complementary strand by hydrogen bonding (non- covalent) between paired bases, adenine (A) with thymine (T) and guanine (G) with cytosine (C). DNA is an abbreviation of DeoxyriboNucleic Acid. : It is the genetic 'blueprint' of all life on this planet. : DNA typing is often applied to Human Identification which has also been called DNA Fingerprinting. DNA Fingerprinting has been attributed to Alex Jeffries, who with his colleagues in 1985, discovered a sequence of nucleotides, subsequently termed short tandem repeats (STR) when comparing the DNA sequence of the human á-globin chromosomes isolated from different individuals.

DNA: Deoxyribonucleic acid, the molecule that stores genetic information. Composed of two complementary strands. See base pairs.

Dominant gene: A gene whose phenotype is when it is present in a single copy.

Dominant(-acting) oncogene: A gene that stimulates cell proliferation and contributes to oncogenesis when present in a single copy.

Dominant: An allele is said to be dominant if it expresses its phenotype even in the presence of a recessive allele. See Allele, Phenotype, Recessive.

Dormancy: A period in which a plant does not grow, awaiting necessary environmental conditions such as temperature, moisture, nutrient availability.

Double helix: Describes the coiling of the antiparallel strands of the DNA molecule, resembling a spiral staircase in which the paired bases form the steps and the sugar-phosphate backbones form the rails.

Double-stranded complementary DNA (dscDNA): A duplex DNA molecule copied from a cDNA template.

Ecosystem: The organisms in a plant population and the biotic and abiotic factors which impact on them. See abiotic factors; Biotic factors.

Electrophoresis: The technique of separating charged molecules in a matrix to which is applied an electrical field.

Electroporation: A method for transforrning DNA, especially useful for plant cells, in which high voltage pulses of electricity are used to open pores in cell membranes, through which foreign DNA can pass.

Electroporation: Transformation technique that uses electric fields to temporarily increase permeability of cells to foreign DNA.

Emergence: A complex whole created by simple parts, as in the brain where billions of neurons work individually, but collectively make up our consciousness and give us the ability to think, rationalize, and create.

Emulation: An absolutely precise simulation of something, so exact that it is equivalent to the original (for example, many computers emulate obsolete computers to run their programs). [AS] The Star Trek replicator is an example.

Enabling science and technologies: Areas of research relevant to a particular goal, such as nanotechnology. Also, technology that "enables" other technology to advance, such as the transistor enabled the computer chip revolution, as did photolithography.

Encapsidation: Process by which a virus' nucleic acid is enclosed in a capsid. See Coat protein.

Endocrine cell: Specialized animal cell that secretes a hormone into the blood; usually part of a gland, such as the thyroid or pituitary gland.

Endocytosis: Uptake of material into a cell by an invagination of the plasma membrane and its internalization in a membrane-bounded vesicle.

Endoplasmic reticulum (ER): Labyrinthine, membrane-bounded compartment in the cytoplasm of eucaryotic cells, where lipids are synthesized and membrane-bound proteins are made.

Endosome: Membrane-bounded organelle in animal cells that carries materials newly ingested by endocytosis and passes many of them on to lysosomes for degradation.

Endothelium: Single sheet of highly flattened cells (endothelial cells) that forms the lining of all blood vessels. Regulates exchanges between the bloodstream and surrounding tissues and is usually surrounded by a basal lamina.

Energy [J]: Capacity for performing work or to cause heat flow. Like work itself, it is measured in Joules.

Environmental Protection Agency (EPA): The U.S. regulatory agency for biotechnology of microbes. The major laws under which the agency has regulatory powers are the Federal Insecticide, Fungicide, and Rodenticide Act (FIFRA); and the Toxic Substances Control Act (TSCA).

Enzymes: Proteins that control the various steps in all chemical reactions.

EPROM: Electrically Programmable Read-Only Memory — nonvolatile memory device.

Escherichia coli: A commensal bacterium inhabiting the human colon that is widely used in biology, both as a simple model of cell biochemical function and as a host for molecular cloning experiments.

Ethidium bromide: A fluorescent dye used to stain DNA and RNA. The dye fluoresces when exposed to UV light.

Eukaryote: An organism whose cells possess a nucleus and other membrane-bound vesicles, including all members of the protist, fungi, plant and animal kingdoms; and excluding viruses, bacteria, and blue-green algae. See Prokaryote.

Eurisko: A computer program developed by Professor Douglas Lenat which is able to apply heuristic rules for performing various tasks, including the invention of new heuristic rules.

Event: An "event" in genetic engineering is the insertion of a particular piece of foreign DNA into the chromosome of the recipient. Insertion occurs in random locations, so each event is unique. The event can affect how a gene is expressed in the organism. Once an event occurs, the transgene can be passed to the next generation as a normally inherited gene.

Evolution: The long-term process through which a population of organisms accumulats genetic changes that enable its members to successfully adapt to environmental conditions and to better exploit food resources.

Exon: A DNA sequence that is ultimately translated into protein. See DNA.

Expectation value: The number of different alignents with scores equivalent to or better than S that are expected to occur in a database search by chance. The lower the E value, the more significant the score.

Express: To translate a gene's message into a molecular product.

Face-centered cubic: A crystal structure found in common elemental metals also FCC. Within the cubic unit cell, atoms are located at all corner and face-centered positions.

Fail-stop: Describes a component or subsystem that, in the event of a failure, produces no output (e.g., of material or data) rather than producing a damaged or incorrect output.

Field-Effect Transistor (FET): Field-Effect Transistor — semiconductor device whose insulated gate electrode controls current flow.

FIFRA: The Federal Insecticide, Fungicide, and Rodenticide Act. See Environmental Protection Agency.

Fingerprint: A set of molecular markers sufficiently diverse to identify particular individuals with reasonable certainty.

Flanking region: The DNA sequences extending on either side of a specific locus or gene.

Food and Drug Administration (FDA): The U.S. agency responsible for regulation of biotechnology food products. The major laws under which the agency has regulatory powers include the Food, Drug, and Cosmetic Act; and the Public Health Service Act.

Fusion gene: A hybrid gene created by joining portions of two different genes (to produce a new protein) or by joining a gene to a different promoter (to alter or regulate gene transcription).

Gamete: A haploid sex cell, egg or sperm, that contains a single copy of each chromosome.

Ganglioside: Any glycolipid having one or more sialic acid residues in its structure. Found in the plasma membrane of eucaryotic cells and especially abundant in nerve cells.

Gel, agarose: A substance used to separate DNA or RNA fragments by size. Used for southern and northern blots.

Gel, polyacrylamide: A substance used to separate proteins by size. Used for western blots.

Gene amplification: The presence of multiple genes. Amplification is one mechanism through which proto-oncogenes are activated in malignant cells.

Gene cloning: The process of synthesizing multiple copies of a particular DNA sequence using a bacteria cell or another organism as a host. See DNA, Host.

Gene expression: The process of producing a protein from its DNA- and mRNA-coding sequences.

Gene flow: The exchange of genes between different but (usually) related populations.

Gene frequency: The percentage of a given allele in a population of organisms. See Allele.

Gene gun: Transformation technique that uses accelerated particles coated with DNA to introduce foreign DNA into recipient plant.

Gene insertion: The addition of one or more copies of a normal gene into a defective chromosome.

Gene linkage: The hereditary association of genes located on the same chromosome.

Gene modification: The chemical repair of a gene's defective DNA sequence. See DNA.

Gene pool: The totality of all alleles of all genes of all individuals in a particular population.

Gene therapy: The introduction of new genes into individuals to cure diseases or genetic abnormalities.

Gene translocation: The movement of a gene fragment from one chromosomal location to another, which often alters or abolishes expression.

Gene: The basic unit of inheritance. A segment of DNA that codes for a particular protein. A locus on a chromosome that encodes a specific protein or several related proteins. It is considered the functional unit of heredity.

Genetic assimilation: Eventual extinction of a natural species as massive pollen flow occurs from another related species and the older crop becomes more like the new crop. See Gene flow.

Genetic code: The three-letter code that translates nucleic acid sequence into protein sequence. The relationships between the nucleotide base-pair triplets of a messenger RNA molecule and the 20 amino acids that are the building blocks of proteins. See Base pair, Nucleic acid, Nucleotide. The genetic information in DNA is encoded with four different nucleotide bases: A, C, G, and T. A set of three consecutive nucleotide bases constitutes a codon. A codon specifies a particular amino acid that is added during synthesis of a protein.

Genetic disease: A disease that has its origin in changes to the genetic material, DNA. Usually refers to diseases that are inherited in a Mendelian fashion, although noninherited forms of cancer also result from DNA mutation.

Genetic drift: Random variation in gene frequency from one generation to another.

Genetic engineering: Genetic engineering is the process of manually adding new DNA on a molecular level with the goal of adding one or more new traits that are not already found in that organism. For the purposes of this web site, genetic engineering includes recombinant DNA and gene splicing technologies.

Genetic linkage map: A linear map of the relative positions of genes along a chromosome. Distances are established by linkage analysis, which determines the frequency at which two gene loci become separated during chromosomal recombination. (

Genetic marker: A gene or group of genes used to "mark" or track the action of microbes.

Genetic modification: The selective and purposeful alteration of genes by humans. Genetic modification includes methods of incorporating new genes or traits into an organism through genetic engineering, mutagenesis and traditional breeding methods.

Genome: The genetic complement contained in the chromosomes of a given organism, usually the haploid chromosome state. The full chromosome set containing all the genes of a particular individual.

Genomic library: A library composed of fragments of genomic DNA.

Genomics: The study of the structure and function of genomes. Genomics usually involves high speed sequencing of the DNA and computer searches for sequences that code for genes.

Genotype: The sum total of the genetic information of an organism including the linkage relationships between genes. The genotype, modified by environment, determines the phenotype.

Genus: A category including closely related species. Interbreeding between organisms within the same category can occur.

Germ cell (germ line) gene therapy: The repair or replacement of a defective gene within the gamete-forming tissues, which produces a heritable change in an organism's genetic constitution.

GMO: Genetically modified organism. An organism that has incorporated a functional foreign gene through recombinant DNA technology. The novel gene exists in all of its cells and is passed through to progeny. Same as transgenic.

Green revolution: Advances in genetics, petrochemicals, and machinery that culminated in a dramatic increase in crop productivity during the third quarter of the 20th century.

Growth factor: A serum protein that stimulates cell division when it binds to its cell-surface receptor.

Growth phase (curve): The characteristic periods in the growth of a bacterial culture, as indicated by the shape of a graph of viable cell number versus time.

GUI: Graphical User Interface — hardware, software, and firmware that produces the display on modern personal computers.

Guy Fawkes Scenario: If nanotechnology becomes widely available, it might become trivial for anyone to committ acts of terrorism (such as making nanomachines build a large amount of explosives under government buildings a la Guy Fawkes). This would either force strict control over nanotechnology (hard) or a decentralized mode of organization.

Hair cell: Specialized sensory epithelial cell in the ear with bundles of giant microvilli (stereocilia) protruding from its apical surface. Sound vibrations tilt the stereocilia, evoking an electrical change in the hair cell, which thus acts as a sound detector.

Hall effect: The phenomenon whereby a force is applied to a moving electron or hole by a magnetic field that is applied perpendicular to the direction of motion. The force direction is perpendicular to both the magnetic field and the particle motion directions.

Haploid cell: A cell containing only one set, or half the usual (diploid) number, of chromosomes.

Heme: An iron complex. Cyclic organic molecule containing an iron atom that carries oxygen in hemoglobin and carries an electron in cytochromes.

Hemophilia: An X-linked recessive genetic disease, caused by a mutation in the gene for clotting factor VIII (hemophilia A) or clotting factor IX (hemophilia B), which leads to abnormal blood clotting.

Herbicide resistance: Herbicide resistance is when plants become resistant to the negative effects of a particular herbicide formulation or are genetically engineered in order to have resistance to a particular herbicide. Herbicide-resistant crops can be sprayed with a particular herbicide, killing weeds growing in and around the crop plants, without being damaged.

Herbicide: Any substance that is toxic to plants; usually used to kill specific unwanted plants.

Heterochromatin: Dark-stained regions of chromosomes thought to be for the most part genetically inactive.

Heterodouplex: A double-stranded DNA molecule or DNA-RNA hybrid, where each strand is of a different origin.

Heterogeneous nuclear RNA (hnRNA): The name originally given to large RNA molecules found in the nucleus, which are now known to be unedited mRNA transcripts, or pre-mRNAs.

Homologous chromosomes: Chromosomes that have the same linear arrangement of genes—a pair of matching chromosomes in a diploid organism. See Chromosomes.

Homologous recombination: The exchange of DNA fragments between two DNA molecules or chromatids of paired chromosomes (during crossing over) at the site of identical nuclcotide sequences.

Homozygote: An organism whose genotype is characterized by two identical alleles of a gene. See Allele, Genotype.

Host: An organism that contains another organism.

Human Genome Project: A project coordinated by the National Institutes of Health (NIH) and the Department of Energy (DOE) to determine the entire nucleotide sequence of the human chromosomes.

Human growth hormone (HGH, somatotrophin): A protein produced in the pituitary gland that stimulates the liver to produce somatomedins, which stimulate growth of bone and muscle.

Hybrid: The offspring of two parents differing in at least one genetic characteristic (trait). Also, a heteroduplex DNA or DNA-RNA molecule.

Hybridization, nucleic acid: Complementary strands of DNA or RNA will spontaneously match up with each other and bond together under the right conditions. This is called nucleic acid hybridization and is used when probing southern and northern blots.

Hybridization: The hydrogen bonding of complementary DNA and/or RNA sequences to form a duplex molecule.

Hybridoma: A hybrid cell, composed of a B Iymphocyte fused to a tumor cell, which grows indefinitely in tissue culture and is selected for the secretion of a specific antibody of interest.

Hydrogen bond: A relatively weak bond formed between a hydrogen atom (which is covalently bound to a nitrogen or oxygen atom) and a nitrogen or oxygen with an unshared electron pair.

Hydrolysis: A reaction in which a molecule of water is added at the site of cleavage of a molecule to two products.

Imaging Contrast Agent: A molecule or molecular complex that increases the intensity of the signal detected by an imaging technique, including MRI and ultrasound. An MRI contrast agent, for example, might contain gadolinium attached to a targeting antibody. The antibody would bind to a specific target—a metastatic melanoma cell, for example – while the gadolinium would increase the magnetic signal detected by the MRI scanner.

Immortalization: Production of a cell line capable of an unlimited number of cell divisions. Can be the result of a chemical or viral transformation or of fusion with cells of a tumor line.

Immortalizing oncogene: A gene that upon transfection enables a primary cell to grow indefinitely in culture.

Incomplete dominance: A condition where a heterozygous off- spring has a phenotype that is distinctly different from, and intermediate to, the parental phenotypes. See Heterozygote, Phenotype.

Initiation codon: The mRNA sequence AUG, coding for methionine, which initiates translation of mRNA.

Inositol lipid: A membrane-anchored phospholipid that transduces hormonal signals by stimulating the release of any of several chemical messengers.

insect resistance management: A set of strategies to reduce the frequency of resistance in crop pests. These strategies are widely referred to as Insect Resistance Management.

Insertion mutations: Changes in the base sequence of a DNA molecule resulting from the random integration of DNA from another source. See DNA, Mutation.

Insulator: A substance, object or material that does not conduct electricity. Material that conducts electricity very poorly.

Insulin: A peptide hormone secreted by the islets of Langerhans of the pancreas that regulates the level of sugar in the blood.

Integrated Circuit (IC): An electronic circuit consisting of many interconnected devices on one piece of semiconductor, typically into 10 millimeters on a side. ICs are the major building blocks of today's computers. Semiconductor circuit, typically on a very small silicon chip, containing microfabricated transistors, diodes, resistors, capacitors, etc.

Interferon: A family of small proteins that stimulate viral resistance in cells.

Intergenic regions: DNA sequences located between genes that comprise a large percentage of the human genome with no known function.

Introgression: Backcrossing of hybrids of two plant populations to introduce new genes into a wild population.

Intron: A noncoding DNA sequence within a gene that is initially transcribed into messenger RNA but is later snipped out. See Coding, DNA, Messenger RNA, Transcription.

Invasiveness: Ability of a plant to spread beyond its introduction site and become established in new locations where it may provide a deliterious effect on organisms already existing there.

Isotope: One of two or more forms of an element that have the same number of protons (atomic number) but differing numbers of neutrons (mass numbers). Radioactive isotopes are commonly used to make DNA probes and metabolic tracers.

Joining (J) segment: A small DNA segment that links genes to yield a functional gene encoding an immunogobulin.

Jupiter-Brain: A posthuman being of extremely high computational power and size. This is the archetypal concentrated intelligence. The term originated due to an idea by Keith Henson that nanomachines could be used to turn the mass of Jupiter into computers running an upgraded version of himself.

Kanamycin: An antibiotic of the aminoglycoside family that poisons translation by binding to the ribosomes.

Karyotype: All of the chromosomes in a cell or an individual organism, visible through a microsope during cell division.

Kb (kilobase): The number of base pairs (in thousands) that denote the size of a nucleic acid fragment.

Kilojoule: Standard unit of energy equal to 1000 joules, or 0.24 kilocalories.

Lag phase: The initial growth phase, during which cell

number remains relatively constant prior to rapid growth. See growth phase.

Lane: On a gel or a blot, a lane belongs to one sample. It contains fragments of different sizes that were separated by gel electrophoresis. See band, blot.

Lawn: A uniform and uninterrupted laver of bacterial growth, in which individual colonies cannot be observed.

Lectin: Protein that binds tightly to a specific sugar. Abundant lectins derived from plant seeds are often used as affinity reagents to purify glycoproteins or to detect them on the surface of cells.

Legume: A member of the pea family that possesses root nodules containing nitrogen-fixing bacteria.

Library, DNA: A large set of clones of DNA fragments from a particular organism. DNA libraries are usually maintained in *E. coli* or in bacteriophages.

Life (lifetime): The length of time the sensor can be used before its performance changes.

Life Science Applications of Deep Red T2-MP EviTag Quantum Dots: The new T2-MP EviTags are available in the deep red wavelengths used in life science research – Macoun Red (660nm) and Jonamac Red (680nm) – with carboxyl, amine, biotin and non-functionalized surfaces. Having a biologically compatible quantum dot label like the T2-MP EviTags allows further exploration and development of in vivo assays, following cells inside living organism to get a new understanding of life. In particular, it offers the possibility to create new classes of assays for use where current testing materials could produce cytotoxicity or cell death. Potential applications include live cell and even whole animal imaging, blood cancer assays, and numerous other applications.

Ligand: In protein chemistry, a small molecule that is (or can be) bound by a larger molecule is termed a ligand. In organometallic chemistry, a moiety bonded to a central metal atom is also termed a ligand; the latter definition is more common in general chemistry.

Ligase (DNA ligase): An enzyme that catalyzes a condensation reaction that links two DNA molecules via the formation of a phosphodiester bond between the 3' hydroxyl and 5' phosphate of adjacent nucleotides.

Light Emitting Diodes (LEDs): Light Emitting Diodes work on a completely different concept. Traditionally LEDs are created from two semiconductors. By running current in one direction across the semiconductor the LED emits light of a particular frequency (hence a particular color) depending on the physical characteristics of the semiconductor used. The semiconductor is covered with a piece of plastic that focuses the light and increases the brightness. These semiconductors are very durable, there is no filament, they donít require much power, theyíre brighter and they last a long time. By densely packing red, blue and green LEDs next to each other on a substrate one can create a display.

Lineage: A chart that traces the flow of genetic information from generation to generation.

Linkage: The frequency of coinheritance of a pair of genes and/or genetic markers, which provides a measure of their physical proximity to one another on a chromosome. Measures the physical distance between two genes. Genes that are close together are unlikely to segregate in a sexual cross. Distant genes segregate independently are are then said to be unlinked.

Linked genes/markers: Genes and/or markers that are

so closely associated on the chromosome that they are coinherited in 80% or more of cases.

Linker: A short, double-stranded oligonucleotide containing a restriction endonuclease recognition site, which is ligated to the ends of a DNA fragment.

Liposome: A type of nanoparticle made of lipids, or fat molecules, surrounding a water core. Liposomes, several of which are widely used to treat infectious diseases and cancer, were the first type of nanoparticle to be used to create therapeutic agents with novel characteristics. Membrane-bound vesicles constructed in the laboratory to transport biological molecules.

Liquid Crystal Display: LCD is the predominant technology used in flat panel displays. The principle that makes the display work is this: A crystalís alignment can be altered with an electric current. If the crystal is lined up one way ñ it will allow the light waves to pass through a polarized filter, but if the electric current alters the crystalís alignment, it will guide light so that the polarized filter blocks the light. By densely packing red, blue and green light emitting crystals next to each other on a sheet (ìcalled a substrateî), one can create a full color display. The great thing about LCD is that the crystals can be packed together closely, allowing for a higher-resolution, finer-detail display. The con is that LCDs are somewhat fragile, require a lot of power and are relatively less bright. Liquid Crystal Display — display device employing light source and electrically alterable optically active thin film.

Locus (plural = loci): A specific location or site on a chromosome.

Logarithmic phase (log or exponential growth phase): The steepest slope of the growth curve—the phase of vigorous growth during which cell number doubles every 20-30 minutes.

Lysogen: A bacterial cell whose chromosome contains integrated viral DNA.

Lysogenic: A type or phase of the virus life cycle during which the virus integrates into the host chromosome of the infected cell, often remaining essentially dormant for some period of time. See Lysogen.

Lytic: A phase of the virus life cycle during which the virus replicates within the host cell, releasing a new generation of viruses when the infected cell lyses.

Machine-phase chemistry: The chemistry of systems in which all potentially reactive moieties follow controlled trajectories (e.g., guided by molecular machines working in vacuum).

Macroscale: Larger than nanoscale; often implies a design that humans can directly interact with; too large to be built by a single assembler (one cubic micron of diamond contains 176 billion atoms).

Magnetization (M): The total magnetic moment per unit volume of material. Also, a measure of the contribution to the magnetic flux by some material within an H field. The magnitude of M is proportional to the applied field as: $M = \div m \times H$, with $\div m$ the magnetic susceptibility.

Magnetostrictive material: A material that changes dimension in the presence of a magnetic field or generates a magnetic field when mechanically deformed.

Malignant: Having the properties of cancerous growth.

Manufacturing: The ability to make products, in this case ranging from clothing, to electronics, to medical devices, to books, to building materials, and much more.

Meat Machine: AKA Cabinet Beast. A box containing assemblers and raw material, within which is formed meat [or whatever else it was programmed to make].

Mechanical: Pertaining to the positions and motions of atoms, as defined by the positions of their nuclei; see electronic. A purely mechanical device can be described in terms of atomic positions and motions without reference to electronic properties, save through their effect on the potential energy function.

Mechanochemistry: Chemistry accomplished by mechanical systems directly controlling the reactant molecules; the formation or breaking of chemical bonds under direct mechanical control. In this volume, the chemistry of processes in which mechanical systems operating with atomic-scale precision either guide, drive, or are driven by chemical transformations. In general usage, the chemistry of processes in which energy is converted from mechanical to chemical form, or vice versa.

Mechanosynthesis: Chemical synthesis controlled by mechanical systems operating with atomic-scale precision, enabling direct positional selection of reaction sites; synthetic applications of mechanochemistry. Suitable mechanical systems include AFM mechanisms, molecular manipulators, and molecular mill systems. Processes that fall outside the intended scope of this definition include reactions guided by the incorporation of reactive moieties into a shared covalent framework (i.e., conventional intramolecular reactions), or by the binding of reagents to enzymes or enzymelike catalysts. Molecular tools with chemically specific tip structures can be used, sequentially, to modify a work piece and build a wide range of molecular structures.

Mechatronics: The study of the melding of AI and electromechanical machines to make machines that are greater than the sum of their parts. The synergistic combination of precision mechanical engineering with electronic control.

Megabase cloning: The cloning of very large DNA fragments.

Megabyte (MB): 220 (= 1,048,576, or about one million) bytes of information.

Meiosis: The reduction division process by which haploid gametes and spores are formed, consisting of a single duplication of the genetic material followed by two mitotic divisions.

Meme: An idea that replicates through a society as it is propagated through person-to-person interaction, both direct and indirect. Memetics is a field of study that focuses on memes' role in the evolution of a culture. Self-reproducing idea or other information pattern which is propagated in ways similar to that of a gene. [Richard Dawkins, 1976]

Messenger RNA (mRNA): The class of RNA molecules that copies the genetic information from DNA, in the nucleus, and carries it to ribosomes, in the cytoplasm, where it is translated into protein.

Metabolism: The biochemical processes that sustain a living cell or organism.

Metallothionein: A protective protein that binds heavy metals, such as cadmium and lead.

Micellar nanocontainers: Block copolymer micelles are water- soluble biocompatible nanocontainers with great potential for delivering hydrophobic drugs. An understanding of their cellular distribution is essential to achieving selective delivery of drugs at the subcellular level. R. Savic, L. Luo, A. Eisenberg, D. Maysinger,

Micellar nanocontainers distribute to defined cytoplasmic organelles Science 300 (5619): 615- 618, Apr. 25, 2003

Micelles: Micelles are small, spherical structures composed of molecules that attract one another to reduce surface tension. The head of the molecule is hydrophilic, meaning it likes water, while the interior portion is hydrophobic, meaning it avoids water.

Microbial mats (biofilms): Layered groups or communities of microbial populations.

Microelectronics: Narrower terms: MEMS, nanoelectronics, optoelectronics, SED Single Electron Devices. Related terms: molecular electronics, semiconductors

Microencapsulation: Individually encapsulated small particles.

Microinjection: A means to introduce a solution of DNA, protein, or other soluble material into a cell using a fine microcapillary pipet.

Microspectrophotometry: Analytical technique for studying substances present at enzyme concentrations in single cells, in situ, by measuring light absorption. Light from a tungsten strip lamp or xenon arc dispersed by a grating monochromator illuminates the optical system of a microscope. The absorbance of light is measured (in nanometers) by comparing the difference between the image of the sample and a reference image. MeSH 1990

Microspheres: Microspheres are generally defined as small spheres made of any material and sized from about 0.5 μm to 100 μm. Similar, but smaller spheres sized 10 to 500 nm are called *nanospheres.* Ideally, microspheres are completely spherical and homogeneous in size, although less perfect particles are often termed microspheres as well. Depending on the preparation method and material used, microspheres show a typical size distribution which often deviates from the mono-sized ideal.

Microstructure: In material engineering the structural features of a material such as grain boundaries, grain size and structure, subject to observation under a microscope, selective etching etc. In MEMS microstructure unfortunately is also used to designate a micromachined feature. The last decade has seen rapid developments in the fabrication, characterization and conceptual understanding of synthetic microstructures in many different material systems including silicon, III-V and II-VI semiconductors, metals, ceramics and organics. The objective of this journal [Superlattices and Microstructures] is to provide a common interdisciplinary platform for the publication of the latest research results on all such "nanostructures" with dimensions in the range of 1-100 nm; the unifying theme here being the dimensions of these artificial structures rather than the material system in which they are fabricated.

Microsystem: A microscale machine that can sense information from the environment and act accordingly. Outside the U.S., it can also refer to microelectromechanical systems (MEMS). [smalltimes glossary, 2002]

MicroTas, micro Total Analysis Systems, uTAS: Although initial research dates back to the early 1970's, the field of micro- TAS formally started in 1990, when Manz et al described the possibility of creating microsystems that would take care of many or all the traditional analytical steps involved in a biochemical analysis (sample introduction, handling, extraction, purification, concentration, filtration, analysis, detection) .... Micro- TAS offer many advantages over traditional analysis systems. Low power consumption and small reaction volumes, faster analysis, ultrasensitive detection, and minimal human intervention are key parameters in the development of micro- TAS. Most biochemical reactions take place in liquid environments. Hence, the development of MicroTAS is intrinsically linked to the design of liquid handling micro- devices. [Biomedical Applications Group (GAB) Centro Nacional de Microelectronica (CNM- IMB) Bellaterra, Spain, 2000]

Microtubule: A hollow cylindrical tube used for the transport of materials inside cells.

Miller indices: A set of 3 integers (4 for hexagonal) that designate crystallographic planes, as determined from reciprocals of fractional axial intercepts.

Millimeter: One thousandth of a meter, or about 1/26 of an inch.

Mitosis: The replication of a cell to form two daughter cells with identical sets of chromosomes.

Molecular - Nanotechnology: The technology of precisely-constructed molecular-scale machines; from nanometer: a billionth of a meter. For a good introduction, see What is Nanotechnology? or Ralph C. Merkle's nanotechnology page.

Molecular assembler: A general-purpose device for molecular manufacturing, able to guide chemical reactions by positioning individual molecules to atomic accuracy (e.g. mechanosynthesis) and to construct a wide range of useful and stable molecular structures according to precise specifications. Also known as an assembler, a molecular assembler is a molecular machine that can build a molecular structure from its component building blocks.

Molecular Beam Epitaxy: [MBE] Process used to make compound (multi-layer) semiconductors. Consists of depositing alternating layers of materials, layer by layer, one type after another (such as the semiconductors gallium arsenide and aluminum gallium arsenide).

Molecular biology: The study of the biochemical and molecular interactions within living cells.

Molecular cloning: The biological amplification of a specific DNA sequence through mitotic division of a host cell into which it has been transformed or transfected.

Molecular genetics: The study of the flow and regulation of genetic information between DNA, RNA, and protein molecules.

Molecular markers: Genetic traits which are detectable on gels or blots and can be used to construct genetic maps. Molecular markers usually have no known function. RFLPs are molecular markers.

Molecular mechanics models: Many of the properties of molecular systems are determined by the molecular potential energy function. Molecular mechanics models approximate this function as a sum of 2-atom, 3-atom, and 4-atom terms, each determined by the geometries and bonds of the component atoms. The 2-atom and 3-atom terms describing bonded interactions roughly correspond to linear springs.

Molecular medicine: A variety of pharmaceutical techniques and gene therapies that address specific molecular diseases or molecular defects in biological systems. A variety of pharmaceutical techniques and therapies in use today. Studying molecules as they relate to health and disease, and manipulating those molecules to improve the diagnosis, prevention, and treatment of disease.

Molecular mill: A mechanochemical processing system characterized by limited motions and repetitive operations without programmable flexibility.

Molecular motors: Protein based machines that are involved in or cause movement such as the rotary devices (flagellar motor and the F1 ATPase) or the devices whose movement is directed along cytoskeletal filaments (myosin, kinesin and dynein motor families). [MeSH, 1999]

Molecular Nanoscience: An emerging interdisciplinary field that combines the study of molecular/ biomolecular systems with the science and technology of nanoscale structures and systems. The potential applications for this research are very broad and include such possibilities as 1) the use of biomolecules and cellular systems to self- assemble nanoelectronic circuitry and other nanoscale structures and 2) the use of lamellar host frameworks containing nano pores that can be tailored to include guest molecules for separation of chemicals for pharmaceutical and other applications. Molecular Nanoscience Alliance for Interdisciplinary Studies and Activities, Univ. of Minnesota, US,

Molecular Wire: A molecular wire—the simplest electronic component—is a quasi-one-dimensional molecule that can transport charge carriers (electrons or holes) between its ends. [Michael D Ward]

Molecularly imprinted polymers MIPs: A new class of materials that have artificially created receptor structures. Since their discovery in 1972, MIPs have attracted considerable interest from scientists and 3engineers involved with the development of chromatographic absorbents, membranes, sensors and enzyme and receptor mimics.

Monoclonal antibodies: Immunoglobulin molecules of single- epitope specificity that are secreted by a clone of B cells.

Monoculture: The agricultural practice of cultivating crops consisting of genetically similar organisms.

Monogenic: Controlled by or associated with a single gene.

Multi-locus probe: A probe that hybridizes to a number of different sites in the genome of an organism.

Mutagen: Any agent or process that can cause mutations. See Mutation.

Mutation: An alteration in DNA structure or sequence of a gene. A spontaneous or induced genetic change in the DNA of an organism. Most mutations are undesirable, but a few are useful such as dwarfing genes in cereal crops that allow the plants to stand better.

Mycorrhizae: Fungi that form symbiotic relationships with roots of more developed plants.

Nanobiology: Many fundamental biological functions are carried out by molecular machineries that have the sizes of 1-100 nm. You find many examples in molecular biology and cell biology: single enzymes, transcription complex, ribosome, transport complex, nuclear pore, and so on. To understand the functions of these machineries, one has to describe their movements, changes in their shapes, and their localization. This means the mechanistic study is equivalent to dynamic morphology at this level of size, making a new field that merges mechanistic biology and morphology. The emergence of nanobiology depended on the invention of nano- technology: scanning probe microscopy, modern optical techniques, and micro- manipulating techniques. This concept of nanobiology was first proposed in a group study named "Biological Nano- Mechanisms", which was supported by Japanese Agency of Science and Technology (1992-1997). [National Institute of Genetics, Japan]

Nanobiotechnology: An emerging area of scientific and technological opportunity. Nanobiotechnology applies the tools and processes of nano/ microfabrication to build devices for studying biosystems. Researchers also learn from biology how to create better micro- nanoscale devices. The Nanobiotechnology Center (NBTC), a National Science Foundation, Science and Technology Center is characterized by its highly interdisciplinary nature and features a close collaboration between life scientists, physical scientists, and engineers. [NBTC, NanoBiotechnology Center, Cornell Univ. US, 2002] Use of nanotechnology[ies] in the life sciences. These may include, but are not limited to, therapeutics, medical devices/implants, biosensors, and tools for the development of drugs. Applying the tools and processes of MNT to build devices for studying biosystems, in order to learn from biology how to create better nanoscale devices. Should hasten the creation of useful micro devices that mimic living biological systems.

Nanocentrifuge: In medical nanorobotics, a proposed nanodevice that can spin materials at very high speed, imparting rotational accelerations of up to one trillion gravities (g's), thus permitting rapid sortation.

Nanoscience: Nanoscience is the science of nanoscale phenomena. This is a very complicated area, since Quantum Mechanics is often needed to describe the physical properties of atoms and molecules properly. What makes it even more complicated is that nanostructured materials can still be quite a challenge to describe Quantum Mechanically, since the equations are feindishly difficult—this is where supercomputers come in to make the job approachable. The scientific discipline seeking to increase our knowledge and understanding of nanoscale phenomena, i.e. science on the scale of 0.1 nm to 100 nm.

Nanosensor: A chemical or physical sensor constructed using nanoscale components, usually microscopic or submicroscopic in size.

Nanosources: sources that emit light from nanometre-scale volumes.

National Institutions of Health (NIH): A nonregulatory agency which has oversight of research activities that the agency funds.

National Science Foundation (NSF): A nonregulatory agency which has oversight of biotechnology research activities that the agency funds.

Natural selection: The differential survival and reproduction of organisms with genetic characteristics that enable them to better utilize environmental resources.

Neophile: One who welcomes the future and who enjoys change and evolution.

Nick translation: A procedure for making a DNA probe in which a DNA fragment is treated with DNase to produce single-stranded nicks, followed by incorporation of radioactive nucleotides from the nicked sites by DNA polymerase I.

Nicked circle (relaxed circle): During extraction of plasmid DNA from the bacterial cell, one strand of the DNA becomes nicked. This relaxes the torsional strain needed to maintain supercoiling, producing the familiar form of plasmid.

Nitrocellulose: A membrane used to immobilize DNA, RNA, or protein, which can then be probed with a labeled sequence or antibody.

Nitrogen fixation: The conversion of atmospheric nitrogen to biologically usable nitrates.

Nitrogenous bases: The purines (adenine and guanine) and pyrimidines (thymine, cytosine, and uracil) that comprise DNA and RNA molecules.

Nodule: The enlargement or swelling on roots of nitrogen-fixing plants. The nodules contain symbiotic nitrogen-fixing bacteria. See Nitrogen fixation.

Nontarget organism: An organism which is affected by an interaction for which it was not the intended recipient.

Northern hybridization: (Northern blotting). A procedure in which RNA fragments are transferred from an agarose gel to a nitrocellulose filter, where the RNA is then hybridized to a radioactive probe.

Novel foods: Novel foods are products that have never been used as a food; foods which result from a process that has not previously been used for food; or, foods that have been modified by genetic manipulation. This last category of foods have been described as genetically modified foods (often referred to as GM foods, genetically engineered foods or biotechnology-derived foods). *Definition from Health Canada Web site*

Nuclease: Any enzyme that cuts nucleic acids. See restriction enzyme. A class of enzymes that degrades DNA and/or RNA molecules by cleaving the phosphodiester bonds that link adjacent nucleotides. In deoxyribonuclease (DNase), the substrate is DNA. In endonuclease, it cleaves at internal sites in the substrate molecule. Exonuclease progressively cleaves from the end of the substrate molecule. In ribonuclease (RNase), the substrate is RNA. In the S1 nuclease, the substrate is single-stranded DNA or RNA.

Nucleic acids: The two nucleic acids, deoxyribonucleic acid (DNA) and ribonucleic acid (RNA), are made up of long chains of molecules called nucleotides. See DNA, RNA, Nucleotides.

Nuclein: The term used by Friedrich Miescher to describe the nuclear material he discovered in 1869, which today is known as DNA.

Nucleoside analog: A synthetic molecule that resembles a naturally occuring nucleoside, but that lacks a bond site needed to link it to an adjacent nucleotide.

Nucleotide: A building block of DNA and RNA, consisting of a nitrogenous base, a five-carbon sugar, and a phosphate group. Together, the nucleotides form codons, which when strung together form genes, which in turn link to form chromosomes.

Nucleus: The membrane-bound region of a eukaryotic cell that contains the chromosomes. A cellular organelle in plants and animals that contains the chromosomes which in turn are composed of DNA plus protein.

Occupational Safety and Health Administration (OSHA): One of the U.S. agencies responsible for regulation of biotechnology. The major law under which the agency has regulatory powers is the Occupational Safety and Health Act.

Oligonucleotide: A DNA polymer composed of only a few nucleotides.

Oncogene: A gene that contributes to cancer formation when mutated or inappropriately expressed.

Oncogenesis: The progression of cytological, genetic, and cellular changes that culminate in a malignant tumor.

Open pollination: Pollination by wind, insects, or other natural mechanisms.

Open reading frame: A long DNA sequence that is uninterrupted by a stop codon and encodes part or all of a protein. (

Operator: A prokaryotic regulatory element that interacts with a repressor to control the transcription of adjacent structural genes.

Orbital: In the approximation that each electron in a molecule has a distinct, independent wave function, the spatial distribution of an electron wave function corresponds to a molecular orbital. These, in turn, can be approximated as sums of contributions from the orbitals characteristic of the isolated atoms. An electron added to a molecule—or, similarly, one excited to a higher-energy state within a molecule—would occupy a state with a different wave function from the rest; an unoccupied state of this kind corresponds to an unoccupied molecular orbital. Orbital-symmetry effects on reaction rates arise when a reaction requires overlap between two lobes of the orbitals on each of two reagents: if the algebraic signs of the wave functions in the facing lobes do not match, bond formation between those orbitals is prohibited.

Organelle: A cell structure that carries out a specialized function in the life of a cell.

Origin of replication: The nucleotide sequence at which DNA synthesis is initiated.

Overlapping reading frames: Start codons in different reading frames generate different polypeptides from the same DNA sequence.

Ovum: A female gamete.

Paleontology: The study of the fossil record of past geological periods and of the phylogenetic relationships between ancient and contemporary plant and animal species.

Palindromic sequence: A DNA locus whose 5'-to-3' sequence is identical on each DNA strand. The sequence is the same when one strand is read left to right and the other strand is read right to left. Recognition sites of many restriction enzymes are palindromic. See DNA.

pAMP: Ampicillin-resistant plasmid developed for this laboratory course.

Parasitism: The closee association of two or more dissimilar organisms where the association is harmful to at least one. See Commensalism, Parasitism, Symbiosis.

Particle bombardment: Using metal particles to blast DNA into cells for the purpose of genetic engineering. A gene gun is used to propel the metal particles.

Pathogen: Organism which can cause disease in another organism.

pBR322: A derivation of ColE1, one of the first plasmid vectors widely used.

PCR: Polymerase chain reaction. An amazingly sensitive technique that allows the specific amplification of extremely small amounts of particular DNA fragments using DNA polymerase and specific primers.

Pedigree: A diagram mapping the genetic history of a particular family.

Percent Accepted Mutation: A unit introduced by Dayhoff et al. to quantify the amount of evolutionary change in a protein sequence. 1.0 PAM unit, is the amount of evolution which will change, on average, 1% of amino acids in a protein sequence. A PAM(x) substitution matrix is a look-up table in which scores for each amino acid substitution have been calculated based on the frequency of that substitution in closely related proteins that have experienced a certain amount (x) of evolutionary divergence.

Persistence: Ability of an organism to remain in a particular setting for a period of time after it is introduced.

Pesticide: A substance that kills harmful organisms (for example, an insecticide or fungicide).

Phenotype: The observable characteristics of an organism, the expression of gene alleles (genotype) as an observable physical or biochemical trait. See Genotype.

Pheromone: A hormone-like substance that is secreted into the environment.

Phosphatase: An enzyme that hydrolyzes esters of phosphoric acid, removing a phosphate group.

Phosphodiester bond: A bond in which a phosphate group joins adjacent carbons through ester linkages. A condensation reaction between adjacent nucleotides results in a phosphodiester bond between 3' and 5' carbons in DNA and RNA.

Phospholipid: A class of lipid molecules in which a phosphate group is linked to glycerol and two fatty acyl groups. A chief component of biological membranes.

Phosphorylation: The addition of a phosphate group to a compound.

Physical map: A map showing physical locations on a DNA molecule, such as restriction sites, and sequence-tagged sites.

Piezoelectric material: A ferroelectric material in which an electrical potential difference is created due to mechanical deformation, or conversely, in which the application of a voltage causes dimensional changes in the material.

Piezoelectric: A piezoelectric material expands or contracts when a voltage is applied across it. Piezoelectric tubes are often used to move the tip in an AFM or STM. If a piezoelectric crystal is compressed, a voltage appears across it. This effect is sometimes used to measure displacements. A commonly used piezoelectric is PZT (Pb,Zr)TiO3.

Plaque: A clear spot on a lawn of bacteria or cultured cells where cells have been Iysed by viral infection.

Plasmid (p): A circular DNA molecule, capable of autonomous replication, which typically carries one or more genes encoding antibiotic resistance proteins. Plasmids can transfer genes between bacteria and are important tools of transformation for genetic engineers.

Pleiotrophy: The effect of a particular gene on several different traits.

Point mutation: A change in a single base pair of a DNA sequence in a gene.

Poly(A) polymerase: Catalyzes the addition of adenine residues to the 3' end of pre-mRNAs to form the poly(A) tail.

Polyacrylamide gel electrophoresis: Electrophoresis through a matrix composed of a synthetic polymer, used to separate proteins, small DNA, or RNA molecules of up to 1000 nucleotides. Used in DNA sequencing.

Polyclonal antibodies: A mixture of immunoglobulin molecules secreted against a specific antigen, each recognizing a different epitope.

Polygenic: Controlled by or associated with more than one gene.

Polylinker: A short DNA sequence containing several restriction enzyme recognition sites that is contained in cloning vectors.

Polymer: A molecule composed of repeated subunits.

Polymerase (DNA): Synthesizes a double-stranded DNA molecule using a primer and DNA as a template.

Polymerase chain reaction (PCR): A procedure that enzymatically amplifies a DNA polymerase.

Polymerase: Any enzyme that adds subunits to chains of macromolecules. Example is DNA polymerase.

Polymorphisms: Variant forms of a particular gene that occur simultaneously in a population.

Polynucleotide: A DNA polymer composed of multiple nucleotides.

Polypeptide (protein): A polymer composed of multiple amino acid units linked by peptide bonds.

Polyploid: A multiple of the haploid chromosome number that results from chromosome replication without nuclear division.

Polysaccharide: A polymer composed of multiple units of monosaccharide (simple sugar).

Polyvalent vaccine: A recombinant organism into which has been cloned antigenic determinants from a number of different disease-causing organisms.

Population: A local group of organisms belonging to the same species and capable of interbreeding.

Primary cell: A cell or cell line taken directly from a living organism, which is not immortalized.

Primer: A short DNA or RNA fragment annealed to single-stranded DNA, from which DNA polymerase extends a new DNA strand to produce a duplex molecule.

Probe: A single-stranded DNA that has been radioactively labeled and is used to identify complementary sequences in genes or DNA fragments of interest. A fragment of DNA, usually labeled with radioactive 32P, that detects homologous sequences on southern blots.

Prokaryote: A bacterial cell lacking a true nucleus; its DNA is usually in one long strand. See Eukaryote.

Promoter: A region of DNA extending 150-300 bp upstream from the transcription start site that contains binding sites for RNA polymerase and a number of proteins that regulate the rate of transcription of the adjacent gene.

Pronucleus: Either of the two haploid gamete nuclei just prior to their fusion in the fertilized ovum.

Protease: An enzyme that cleaves peptide bonds that link amino acids in protein molecules.

Protein: Molecules composed of amino acids. Proteins constitute the enzymes and many of the structural components of cells. A polymer of amino acids linked via peptide bonds and which may be composed of two or more polypeptide chains. Proteins do the work in cells. They can be part of structures (such as cell walls, organelles, etc), regulate reactions that take place in the cell or they can serve as enzymes, which speed-up reactions. Everything you see in an organism is either made of proteins or the result of a protein action.

Protein kinase: An enzyme that adds phosphate groups to a protein molecule at serine, threonine, or tyrosine residues.

Proteinaceous infectious particle (prion): A proposed pathogen composed only of protein with no detectable nucleic acid and which is responsible for Creutzfeldt-Jakob disease and kuru in humans and scrapie in sheep.

Quantum Mechanics: A largely computational physical theory explaining the behavior of quantum phenomena, which incorporates the theory of special relativity. Despite dilignet attempts, general relativity has not been sucessfully incorporated into quantum mechanics. A physical model of chemical and optical phenomena, as well as the behaviour of matter in general on a small scale. Quantum mechanics describes a system of particles in terms of a wave function defined over the configuration space of the system. Although the concept of particles having distinct locations is implicit in the potential energy function that determines the wave function (e.g., of a ground-state system), the observable dynamics of the system cannot be described in terms of the motion of such particles from point to point. In describing the energies, distributions, and behaviors of electrons in nanometer-scale structures, quantum mechanical methods are necessary. Electron wave functions help determine the potential energy surface of a molecular system, which in turn is the basis for classical descriptions of molecular motion. Nanomechanical systems can almost always be described in terms of classical mechanics, with occasional quantum mechanical corrections applied within the framework of a classical model.

Quantum Well: A P-N-P junction in which the "N" layer is ~10 nm (where traditional physics leaves off and quantum effects take over) and an "electron trap" is created. "If one makes a heterostructure with sufficiently thin layers, quantum interference effects begin to appear prominently in the motion of the electrons. The simplest structure in which these may be observed is a quantum well, which simply consists of a thin layer of a narrower-gap semiconductor between thicker layers of a wider-gap material."

Reading frame: A series of triplet codons beginning from a specific nucleotide. Depending on where one begins, each DNA strand contains three different reading frames.

Recessive gene: Characterized as having a phenotype expressed only when both copies of the gene are mutated or missing.

Recessive(-acting) oncogene, (anti-oncogene): A single copy of this gene is sufficient to suppress cell proliferation; the loss of both copies of the gene contributes to cancer formation.

Recessive: Refers to the member of a pair of alleles that fails to be expressed in the phenotype of the organism when the dominant member is present. Also refers to the phenotype of an individual that has only the recessive allele.

Recognition sequence (site): A nucleotide sequence—composed typically of 4, 6, or 8 nucleotides—that is recognized by a restriction endonuclease. Type II enzyrnes cut (and their corresponding modification enzymes methylate) within or very near the recognition sequence.

Recombinant DNA: The process of cutting and recombining DNA fragments from different sources as a means to isolate genes or to alter their structure and function.

Recombination frequency: The frequency at which crossing over occurs between two chromosomal loci—the probability that two loci will become unlinked during meiosis.

Refuge: A refuge is any host plant (non-Bt crop, potatoes, oats, sorghum, and some weeds) not producing Bt proteins or not being treated with conventional Bt formulations. The purpose of the refuge is to supply a source of susceptible insects that may mate with resistant insects of the same species emerging from nearby Bt crop fields. If the refuge is large enough, susceptible insect numbers should outnumber resistant individuals, and in theory, a greater number of the next generation offspring will be susceptible to the Bt crop.

Regulatory gene: A gene whose protein controls the activity of other genes or metabolic pathways.

Relaxed plasmid: A plasmid that replicates independently of the main bacterial chromosome and is present in 10-500 copies per cell.

Renature: The reannealing (hydrogen bonding) of single-stranded DNA and/or RNA to form a duplex molecule.

Replicon: A chromosomal region containing the DNA sequences necessary to initiate DNA replication processes.

Repressor: A DNA-binding protein in prokaryotes that blocks gene transcription by binding to the operator.

Resistance: The evolved ability of a species to survive during a stressful event or set of circumstances.

Restriction endonuclease (enzyme): A class of endonucleases that cleaves DNA after recognizing a specific sequence, such as BamH1 (GGATCC), EcoRI (GAATTC), and HindIII (AAGCTT). Type I. Cuts nonspecifically a distance greater than 1000 bp from its recognition sequence and contains both restriction and methylation activities. Type II. Cuts at or near a short, and often symmetrical, recognition sequence. A separate enzyme methylates the same recognition sequence. Type III. Cuts 24-26 bp downstream from a short, asymmetrical recognition sequence. Requires ATP and contains both restriction and methylation activities.

Restriction enzyme: A class of enzymes that cut DNA at specific sequences called restriction sites. Restriction enzymes were key to making genetic engineering possible. See RFLP.

Restriction-fragment-length polymorphism (RFLP): Differences in nucleotide sequence between alleles at a chromosomal locus result in restriction fragments of varying lengths detected by Southern analysis.

Retrovirus: A member of a class of RNA viruses that utilizes the enzyme reverse transcriptase to reverse copy its genome into a DNA intermediate, which integrates into the hostcell chromosome. Many naturally occurring cancers of vertebrate animals are caused by retroviruses.

Reverse genetics: Using linkage analysis and polymorphic markers to isolate a disease gene in the absence of a known metabolic defect, then using the DNA sequence of the cloned gene to predict the amino acid sequence of its encoded protein.

Reverse transcriptase (RNA-dependent DNA polymerase): An enzyme isolated from retrovirus-infected cells that synthesizes a complementary (c)DNA strand from an RNA template.

RFLP: Restriction fragment length polymorphism. RFLPs are generated by digesting DNA with restriction enzymes. The DNA is separated by size by gel electrophoresis. Slight differences in homologous fragments may exist between individuals. These length polymorphisms are RFLPs. See blot.

Rhizobia: Bacteria in a symbiotic relationship with leguminous plants that results in nitrogen fixation. See Nitrogen fixation.

Rhizosphere: The soils region on and around plant roots.

Ribosomal RNA (rRNA): The RNA component of the ribosome.

Ribosome: Cellular organelle that is the site of protein synthesis during translation. See Organelle, Translation.

Ribosome-binding site: The region of an mRNA molecule that binds the ribosome to initiate translation.

Risk assessment: Defined as a formalized basis for the objective evaluation of risk in a manner in which assumptions and uncertainties are clearly considered and presented.

RNA (ribonucleic acid): An organic acid composed of repeating nucleotide units of adenine, guanine, cytosine, and uracil, whose ribose components are linked by phosphodiester bonds.

RNA polymerase: Transcribes RNA from a DNA template.

RNA: Ribonucleic acid. The three basic types are messenger RNA (mRNA), ribosomal RNA (rRNA), and transfer RNA (tRNA).

Roundup Ready: A trademark for plants that are genetically engineered to be resistant to the herbicide Roundup (technical name: glyphosate).

Salmonella: A genus of rod-shaped, gram-negative bacteria that are a common cause of food poisoning.

Satellite RNA (viroids): A small, self-splicing RNA molecule that accompanies several plant viruses, including tobacco ringspot virus.

Scanning Near Field Optical Microscopy: A method for observing local optical properties of a surface that can be smaller than the wavelength of the light used. A type of scanning probe microscopy that maps the near-field optical properties of an interface.

Scanning Probe Microscopy (SPM): A method for imaging

nanoscale features of surfaces by scanning a sensor (probe) over a surface. Near-field effects such as tunneling, van der Waals forces, local fields and more are serially detected at localized points on the surface and used to create an SPM image.

Scanning Thermal Microscopy (SThM): A type of scanning probe microscopy that maps the local temperature and thermal conductivity of an interface. A method for observing local temperatures and temperature gradients on a surface. [NTN]

Scanning Tunneling Microscope (STM): An instrument able to image conducting surfaces to atomic accuracy; has been used to pin: molecules to a surface. A device in which a sharp conductive tip is moved across a conductive surface close enough to permit a substantial tunneling current (typically a nanometer or less). In a common mode of operation, the voltage is kept constant and the current is monitored and kept constant by controlling the height of the tip above the surface; the result, under favorable conditions, is an atomic-resolution map of the surface reflecting a combination of topography and electronic properties. The STM has been used to manipulate atoms and molecules on surfaces. A type of scanning probe microscopy that maps the local electrical properties of an interface.

Schwann cell: Glial cell responsible for forming myelin sheaths in the peripheral nervous system.

Science: The process of developing a systematized knowledge of the world through the variation and testing of hypotheses. The pursuit of knowledge and understanding, from the Latin term *scientia*, which means 'knowledge'.

Selectable marker genes: Genetic engineers must be

able to select cells that have received the transgene from those that have not. Selecting out transgenic cells is done by co-transforming the cells with the transgene plus an additional gene called a selectable marker gene. Selectable marker genes are genes that encode easily detectable traits making transgenic cells discernible from non-transgenic cells. The two most commonly used selectable marker genes encode the traits of herbicide and antibiotic resistance.

Selectable marker: A gene whose expression allows one to identify cells that have been transforrned or transfected with a vector containing the marker gene.

Self-pollination: Pollen of one plant is transferred to the female part of the same plant or another plant with the same genetic makeup.

Semiconservative replication: During DNA duplication, each strand of a parent DNA molecule is a template for the synthesis of its new complementary strand. Thus, one half of a preexisting DNA molecule is conserved during each round of replication.

Sequence hypothesis: Francis Crick's seminal concept that genetic information exists as a linear DNA code; DNA and protein sequence are colinear.

Sequence: Noun: the particular order of nucleotides in a DNA or RNA fragment. Verb: to determine the particular order of nucleotides in a strand of DNA.

Sequence-tagged site (STS): A unique (single-copy) DNA sequence used as a mapping landmark on a chromosome.

Sexual reproduction: The process where two cells (gametes) fuse to form one hybrid, fertilized cell. See Asexual reproduction, Gamete, Hybrid.

Signal transduction: The biochemical events that conduct the signal of a hormone or growth factor from the cell exterior, through the cell membrane, and into the cytoplasm. This involves a number of molecules, including receptors, pro- teins, and messengers.

Site-directed mutagenesis: The process of introducing spe- cific base-pair mutations into a gene.

Small nuclear RNA (snRNA): Short RNA transcripts of 100-300 bp that associate with proteins to form small nuclear ribonucleoprotein particles (snRNPs), which participate in RNA processing.

Somatic cell gene therapy: The repair or replacement of a defective gene within somatic tissue.

Somatic cell: Any nongerm cell that composes the body of an organism and which possesses a set of multiploid chromosomes (diploid in most organisms).

Southern hybridization (Southern blotting): A procedure in which DNA restriction fragments are transferred from an agarose gel to a nitrocellulose filter, where the denatured DNA is then hybridized to a radioactive probe (blotting).

Species: A classification of related organisms that can freely interbreed.

Spore: A form taken by certain microbes that enables them to exist in a dormant stage. It is an asexual reproductive cell. See Asexual reproduction, Dormant.

Stationary phase: The plateau of the growth curve after log growth, during which cell number remains constant. New cells are produced at the same rate as older cells die.

Sticky end: A protruding, single-stranded nucleotide se- quence produced when a restriction endonuclease cleaves off center in its recognition sequence. After cutting with restriction enzymes, the ends of DNA fragments are "sticky". They can easily be joined with other fragments that were cut with the same enzyme and so have complementary sticky ends.

Stringency: Reaction conditions—notably temperature, salt, and pH—that dictate the annealing of single-stranded DNA/DNA, DNA/RNA, and RNA/RNA hybrids. At high stringency, duplexes form only between strands with perfect one-to-one complementarity; lower stringency allows annealing between strands with some degree of mismatch between bases.

Stringent plasmid: A plasmid that only replicates along with the main bacterial chromosome and is present as a single copy, or at most several copies, per cell.

Structure-functionalism: The scientific tradition that stresses the relationship between a physical structure and its function, for example, the related disciplines of anatomy and physiology.

Subcloning: The process of tranferring a cloned DNA fragment from one vector to another.

Substantial equivalence: The concept of substantial equivalence is used as a guide in the safety assessment of genetically modified foods by comparing the novel food to its unmodified counterpart which has a history of safe use. This approach allows regulatory agencies to include in their consideration, the substantial history of information related to foods which have long been safely consumed in the human diet to aid in the identification of potential safety and nutritional issues. *Definition from Health Canada Web site.*

Subunit vaccine: A vaccine composed of a purified antigenic determinant that is separated from the virulent organism.

Supercoiled plasmid: The predominant in vivo form of

plasmid, in which the plasmid is coiled around histone-like proteins. Supporting proteins are stripped away during extraction from the bacterial cell, causing the plasmid molecule to supercoil around itself in vitro.

Supergene: A group of neighboring genes on a chromosome that tend to be inherited together and sometimes are functionally related.

Supernatant: The soluble liquid &action of a sample after centrifugation or precipitation of insoluble solids.

Susceptibility: When individual or species will become ill or possibly die in the face of the stressful event or circumstances.

Symbiosis: The close association of two or more dissimilar organisms where both receive an advantage from the association. See Commensalism, Parasitism.

Synapsis: The pairing of homologous chromosome pairs during prophase of the first meiotic division, when crossing over occurs.

Taq polymerase: A heat-stable DNA polymerase isolated from the bacterium Therrnus aquaticus, used in PCR.

TATA box: An adenine- and thymine-rich promoter sequence located 25-30 bp upstream of a gene, which is the binding site of RNA polymerase.

T-DNA (transfer DNA, tumor-DNA): The transforming region of DNA in the Ti plasmid of Agrobacterium tumefaciens.

Template: An RNA or single-stranded DNA molecule upon which a complementary nucleotide strand is synthesized.

Termination codon: Any of three mRNA sequences (UGA, UAG, UAA) that do not code for an amino acid and thus signal the end of protein synthesis. Also known as stop codon.

Terminator region: A DNA sequence that signals the end of transcription.

Tesla [T]: Unit of magnetic induction: 1T = 1 weber/m2 (also, 1T = 104 gauss).

Thermistor: A temperature-measuring device, that contains a resistor or semiconductor whose resistance varies with temperature.

Thermocouple: A temperature-measuring device, which contains a pair of end-joined dissimilar conductors in which an electromotive force is developed by thermoelectric effects when the joined ends and the free ends of the conductors are a different temperature.

Thiol: An SH group, or a molecule containing one. Also known as a sulfhydryl or mercapto group.

Threshold: The smallest input signal that will cause a readable change in the output signal.

Thrombin : The enzyme that converts *fibrinogen* to *fibrin* to form a blood clot. Thrombin circulates in the inactive form, prothrombin.

Thylakoid: Flattened sac of membrane in a chloroplast that contains pigment and carries out the light-gathering reactions of photosynthesis. Stacks of thylakoids form the grana of chloroplasts.

Thymidine kinase (tk): An enzyme that allows a cell to utilize an alternate metabolic pathway for incorporating thymidine into DNA. Used as a selectable marker to identify transfected eukaryotic cells.

Ti (tumor-inducing) plasmid: A giant plasmid of Agrobacterium tumefaciens that is responsible for tumor formation in infected plants. Ti plasmids are used as vectors to introduce foreign DNA into plant cells.

Toxic Substances Control Act (TSCA): See Environmental Protection Agency.

Transcapsidation: The partial of full coating of the nucleic acid of one virus with a coat protein of a differing virus. See Coat protein.

Transcription: The process of creating a complementary RNA copy of DNA.

Tanscription: The copying of genetic information from DNA to messenger RNA (mRNA).

Transduction: The transfer of DNA sequences from one bacterium to another via lysogenic infection by a bacteriophage (transducing phage).

Transfection: The uptake and expression of a foreign DNA sequence by cultured eukaryotic cells.

Transformant: In prokaryotes, a cell that has been genetically altered through the uptake of foreign DNA. In higher eukaryotes, a cultured cell that has acquired a malignant phenotype.

Transformation: (1) Transformation is the step in the genetic engineering process where a new gene (transgene) is delivered into the nucleus of a plant cell and inserts into a chromosome where it is passed on to progeny. Transformation means to genetically change a living thing. (2) In prokaryotes, the natural or induced uptake and expression of a foreign DNA sequence—typically a recombinant plasmid in experimental systems. In higher eukaryotes, the conversion of cultured cells to a malignant phenotype—typically through infection by a tumor virus or transfection with an oncogene.

Transformation efficiency: The number of bacterial cells that uptake and express plasmid DNA divided by the mass of plasmid used (in transformants/microgram).

Transforming oncogene: A gene that upon transfection converts a previously immortalized cell to the malignant phenotype.

Transgene: A foreign gene incorporated by transformation.

Transgenic: A genetically engineered plant created through transformation is sometimes referred to as a transgenic.

Transgenic animal: Genetically enginnered animal or offspring of genetically engineered animals. The transgenic animal usually contains material from at lease one unrelated organism, such as from a virus, plant, or other animal. See Transgenic.

Transgenic plant: Genetically engineered plant or offspring of genetically engineered plants. The transgenic plant usually contains material from at least one unrelated organisms, such as from a virus, animal, or other plant. See Transgenic.

Transgenic: An organism in which a foreign DNA gene (a transgene) is incorporated into its genome early in development. The transgene is present in both somatic and germ cells, is expressed in one or more tissues, and is inherited by offspring in a Mendelian fashion. See Transgenic animal, Transgenic plant.

Transgenic: An organism that has incorporated a functional foreign gene through recombinant DNA technology. The novel gene exists in all of its cells and is passed through to progeny. Same as genetically modified organism (GMO).

TSCA: The Toxic Substances Control Act. See Environmental Protection Agency.

Tumor virus: A virus capable of transforming a cell to a malignant phenotype.

U:S: Department of Agriculture: The U.S. agency

responsible for regulation of biotechnology products in plants and animals. The major laws under which the agency has regulatory powers include the Federal Plant Pest Act (PPA), the Federal Seed Act, and the Plant Variety Act (PVA). In addition, the Science and Education (S&E) division has nonregulatory oversight of research activities that the agency funds.

Vaccinia: The cowpox virus used to vaccinate against smallpox and, experimentally, as a carrier of genes for antigenic determinants cloned from other disease organisms.

Variable surface glycoprotein (VSG): One of a battery of antigenic determinants expressed by a microorganism to elude immune detection.

Variation: Differences in the frequency of genes and traits among individual organisms within a population.

Vasculocyte: In medical nanorobotics, a theorized (nanorobotic) device capable of performing repairs of an injured vascular luminal surface.

Vasculography: A physical description (and mapping) of the human vascular system.

VC (Venture capitalist): A person or organisation that invests in promising companies during early development stages.

Viral oncogene: A viral gene that contributes to malignancies in vertebrate hosts.

Viroid: A plant pathogen that consists of a naked RNA molecule of approximately 250-350 nucleotides, whose extensive base pairing results in a nearly correct double helix.

Virtual Nanomedicine: Using VR to perform surgery and other functions inside the body.

Virtual reality system: A combination of computer and interface devices (goggles, gloves, etc.) that presents a user with the illusion of being in a three dimensional world of computer-generated objects.

Virus: A parasite (consisting primarily of genetic material) that invades cells and takes over their molecular machinery in order to copy itself. A small replicator consisting of little but a package of DNA or RNA which, when injected into a host cell, can direct the cell's molecular machinery to make more viruses. Particle consisting of nucleic acid (RNA or DNA) enclosed in a protein coat and capable of replicating within a host cell and spreading from cell to cell. Often the cause of disease.

1

Biotechnology Science: An Introductory Overview of Basics

Biotechnology has many applications in health, agriculture, aquaculture and the environment. Health Canada's role in biotechnology is to protect the health and safety of Canadians through regulation and public health policy. Biotechnology has been used by humans for thousands of years. For example, biotechnology has been used to make cheese, ferment wine and beer and make bread by using micro-organisms such as bacteria or fungi. Over time, we have also domesticated and selectively bred some animals and plants to meet human needs. In Canada, Health Canada, the Canadian Food Inspection Agency, Fisheries and Oceans Canada and Environment Canada, have joint responsibility to regulate biotechnology-derived products. The definition used is found in the Canadian Environmental Protection Act (1999): "the application of science and engineering to the direct or indirect use of living organisms of parts or products of living organisms, in their natural or modified forms." Examples of biotechnology-derived products regulated by Health Canada are genetically modified and other novel foods, biologics and genetic therapies and assisted human reproduction technologies.

Biotechnology Techniques

Biotechnology can be best understood as an umbrella

term referring to a variety of techniques, both traditional and modern. Note that some traditional techniques, such as selective breeding, hybridization and mutagenesis, are used in current applications of biotechnology, only with increased scientific knowledge and more advanced technology. These include:

- Fermentation: The breakdown of complex organic substances by micro-organisms, in the absence of oxygen. The process is energy-yielding.
- Selective breeding: The breeding of selected plants and animals to produce offspring with desired traits. The offspring with the desired traits are then used as breeding stock for the next generation and so on, until offspring that express the desired traits are obtained.
- Hybridization: The production of offspring, known as hybrids, from genetically dissimilar parents. The object of hybridization is to combine desirable genes found in two or more different varieties to produce pure-breeding offspring superior in many respects to the parental types.
- Mutagenesis: The use of mutagens (such as exposure to radiation, temperature extremes and certain chemicals), to cause changes in the genetic make-up of cells, possibly resulting in new desirable, inheritable traits.
- Recombinant DNA (rDNA) techniques: The application of genetic techniques to produce desirable traits in living organisms using other living organisms such as bacteria. Examples of these techniques include the use of: restriction endonucleases (enzymes, produced by bacteria, that break foreign DNA molecules with the gene of interest

into fragments which recombine with complementary molecules from a different source to form a recombinant DNA molecule); vectors (plasmids, often bacterial, or viruses that carry a piece of DNA into a bacterium for cloning purposes); and gene guns (called DNA particle guns and used to fire genes into cells by coating the genes onto tiny gold or tungsten particles and firing them from the barrel of the gun).

- Tissue Culture: A biological technique in which fragments of plant or animal tissue or cells are transferred to an artificial environment, free of other organisms, in which they continue to survive and function for reproduction, chemical production and medical research.

Bioinformatics

Major changes in the field of biology have taken place over the last few decades. Advances in biotechnology and accomplishments such as the mapping of the entire human genome and the genomes of many other organisms are producing a better understanding of biological systems and living organisms. Large amounts of biological information and data have been and continue to be generated through this research. Bioinformatics arose from the need to organize, store, and analyse this information.

Bioinformatics is the use of computer software and programs to organize, store and analyse biological information and data to better understand biological systems. It is the field of science where biology, computer science and information technology converge as a single discipline. Researchers use advanced computer and statistical techniques to wade through and analyse large amounts of biological data created through modern biotechnologies referred to as the "omics" technologies – such as genomics and proteomics. These "omics" technologies focus on different aspects of biological systems. For example:

- Genomics is the analysis of the entire genetic make-up of an organism. It includes a study of the functions of genes in cells, organs and organisms.
- Proteomics focusses on the total protein make-up of a cell at any given time.

These "omics" technologies generate massive amounts of biological data through which it would be impossible to navigate without the use of computer systems. The data includes sequences of amino acids and nucleotides that underlie genes and proteins. The goal of bioinformatics is to translate the complex data gathered into usable knowledge. Bioinformatics typically includes three main areas:

- Developing new statistics and data to be studied and used to assess relationships among the different sets of biological data.
- Analysing and interpreting the various types of data, which include nucleotide and amino acid sequences and the properties of various proteins.
- Developing and implementing tools to enable efficient access and management of different types of pertinent information.

The field of bioinformatics emerged in the early 1980s with the creation of the GenBank database. GenBank, started by the US Department of Energy, stores DNA sequence information obtained from a wide range of organisms. In the early days, GenBank was a small scale operation with a roomful of technicians sitting at keyboards entering the DNA sequence information published in academic journals.

The advent of the Internet allowed researchers to access data in GenBank from all over the world for free. The DNA sequence data in GenBank grew rapidly with the emergence of highly sophisticated gene sequencing tools. Private companies joining the sequencing race with parallel projects created huge databases of their own.

Several services emerged as the need for access to bioinformatics increased. The two most significant were the European Molecular Biology Network (EMBnet) and the United States National Center for Biotechnology Information (NCBIO). In Canada, the Canada Institute for Scientific and Technical Information (CISTI) of the National Research Council of Canada (NRC) operated a bioinformatics server, Molecular Biology Database Service (MBDS), from 1987 to 1996. The Canadian Bioinformatics Resource (CBR) was created in 1995 to replace MBDS in providing bioinformatics resources and services to Canadian researchers.

Bioinformatics Applications

Most bioinformatics tools currently available deal with the structural and functional aspects of genes and proteins – looking at what genes and proteins look like, where they are located and what they do. Applications of bioinformatics allow researchers to accomplish the following:

- Compare mapped genomes of any species to identify similarities and differences among organisms.
- Researchers can access genetic information from around the world. Various searchable databases, many of which are on the Internet, are used to identify DNA sequences of identified genes.
- Databases also allow quick access to information about proteins. Researchers retrieve a sequence using

a protein name, or retrieve a protein using the amino acid sequence.

- Sequence translation programs enable users to analyse nucleotide sequences. The programs convert the DNA sequences into protein sequences or protein sequences into their complementary DNA (cDNA).
- Sequencing tools are used by bioinformatists to find out more about the characteristics of genetic material, including proteins. Once the protein sequences are analysed, the researchers can use software programs to search databases to find similar sequences. Other programs allow researchers to look at the three-dimensional shape of the proteins and nucleotides. Since shape is a critical determinant of function, these programs allow researchers to gain clues for identifying the roles of the sequences.

Examples of Common Bioinformatics Tools

Data produced through research from all over the world is collected and organized in databases specialized in individual subjects. Computational tools are used to efficiently analyse this data. Many areas of bioinformatics focus on predicting the biological functions of genes, proteins, and their parts based on the structural data compiled. Examples of commonly used tools include:

- BLAST (Basic Local Alignment Search Tool) — This tool allows researchers to identify similarities between their own nucleotide or protein sequence with those in the public databases. It is used to identify whether a given sequence is new, similar to a known sequence (e.g. other proteins), or contains specific protein characteristics which may give a clue to the possible role of the sequence.

- OMIM (Online Mendelian Inheritance in Man) — This database contains information about human genes and genetic disorders already mapped in the human genome. Typically, the information contained includes a summary of each disease, including symptoms, genetic mapping information, related references, and a direct link to the DNA sequence in GenBank if the gene responsible for the disease has been cloned. It also contains mapping information on genes not yet linked to diseases.

GeneMap98 — This database lists genes mapped in the human genome. It is helpful to researchers who are looking to identify genes for diseases with simple traits or symptoms since the map can display genes within a location defined by the user.

Bioinformatics is an interdisciplinary field that addresses biological problems using computational techniques. The field is also often referred to as computational biology. It plays a key role in various areas like functional genomics, structural genomics, and proteomics amongst others, and forms a key component in biotechnology and pharmaceutical sector. Bioinformatics and computational biology involve the use of techniques from applied mathematics, informatics, statistics, and computer science to solve biological problems. Research in computational biology often overlaps with systems biology. Major research efforts in the field include sequence alignment, gene finding, genome assembly, protein structure alignment, protein structure prediction, prediction of gene expression and protein-protein interactions, and the modeling of evolution. The terms *bioinformatics* and *computational biology* are often used interchangeably, although the former typically focuses on algorithm development and specific computational methods, while the latter focuses more on hypothesis testing and discovery in the biological domain. Although this distinction is used by National Institutes of Health in their working definitions of Bioinformatics and Computational Biology, it is clear that there is a tight coupling of developments and knowledge between the more hypothesis-driven research in computational biology and technique-driven research in bioinformatics. Computational biology also includes lesser-known but equally important sub-disciplines such as computational biochemistry and computational biophysics. A common thread in projects in bioinformatics and computational biology is the use of mathematical tools to extract useful information from noisy data produced by high-throughput biological techniques such as genomics (The field of data mining overlaps with computational biology in this regard). A representative problem in bioinformatics is the assembly of high-quality DNA sequences from fragmentary "shotgun" DNA sequencing, while in computational biology, a representative problem might be statistical testing of a hypothesis of common gene regulation using data from mRNA microarrays or mass spectrometry.

Sequence analysis

Since the Phage Ö-X174 was sequenced in 1977, the DNA sequences of more and more organisms have been decoded and stored in electronic databases. This data is analyzed to determine genes that code for proteins, as well as regulatory sequences. A comparison of genes within a species or between different species can show similarities between protein functions, or relations between species (the use of molecular systematics to construct phylogenetic trees). With the growing amount of data, it long ago became impractical to analyze DNA sequences manually. Today, computer programs are used to search the genome of thousands of organisms, containing billions of nucleotides. These programs can compensate for mutations (exchanged, deleted or inserted bases) in the DNA sequence, in order to identify sequences that are related, but not identical. A variant of this sequence alignment is used in the sequencing process itself. The so-called shotgun sequencing technique (which was used, for example, by The Institute for Genomic Research to sequence the first bacterial genome, *Haemophilus influenza*) does not give a sequential list of nucleotides, but instead the sequences of thousands of small DNA fragments (each about 600-800 nucleotides long). The ends of these fragments overlap and, when aligned in the right way, make up the complete genome. Shotgun sequencing yields sequence data quickly, but the task of assembling the fragments can be quite complicated for larger genomes. In the case of the Human Genome Project, it took several months of CPU time (on a circa-2000 vintage DEC Alpha computer) to assemble the fragments. Shotgun sequencing is the method of choice for virtually all genomes sequenced today, and genome assembly algorithms are a critical area of bioinformatics research. Another aspect of bioinformatics in sequence analysis is the automatic search for genes and regulatory sequences within a genome. Not all of the nucleotides within a genome are genes. Within the genome of higher organisms, large parts of the DNA do not serve any obvious purpose. This so-called junk DNA may, however, contain unrecognized functional elements. Bioinformatics helps to bridge the gap between genome and proteome projects, for example in the use of DNA sequence for protein identification.

Genome annotation

In the context of genomics, annotation is the process of marking the genes and other biological features in a DNA sequence. The first genome annotation software system was designed in 1995 by Owen White, who was part of the team that sequenced and analyzed the first genome of a free-living organism to be decoded, the bacterium Haemophilus influenzae. Dr. White built a software system to find the genes (places in the DNA sequence that encode a protein), the transfer RNA, and other features, and to make initial assignments of function to those genes. Most current genome annotation systems work similarly, but the programs available for analysis of genomic DNA are constantly changing and improving.

Computational evolutionary biology

Evolutionary biology is the study of the origin and descent of species, as well as their change over time. Informatics has assisted evolutionary biologists in several key ways; it has enabled researchers to:

- trace the evolution of a large number of organisms by measuring changes in their DNA, rather than through physical taxonomy or physiological observations alone,
- more recently, compare entire genomes, which permits the study of more complex evolutionary events, such as gene duplication, lateral gene transfer, and the prediction of bacterial speciation factors,
- build complex computational models of populations to predict the outcome of the system over time
- track and share information on an increasingly large number of species and organisms

Future work endeavors to reconstruct the now more complex tree of life. The area of research within computer science that uses genetic algorithms is sometimes confused with computational evolutionary biology. Work in this area involves using specialized computer software to improve equations, algorithms, or integrated circuit designs. It is inspired by evolutionary principles such as replication, diversification through recombination or mutation, fitness, survival through selection or culling, and iteration, collectively called a Darwinian machine or Darwinian ratchet.

Measuring biodiversity

Biodiversity of an ecosystem might be defined as the total genomic complement of a particular environment, from all of the species present, whether it is a biofilm in an abandoned mine, a drop of sea water, a scoop of soil, or the entire biosphere of the planet Earth. Databases are used to collect the species names, descriptions, distributions, genetic information, status and size of populations, habitat needs, and how each organism interacts with other species. Specialized software programs are used to find, visualize, and analyze the information, and most importantly, communicate it to other people. Computer simulations model such things as population dynamics, or calculate the cumulative genetic health of a breeding pool (in agriculture) or endangered population (in conservation). One very exciting potential of this field is that entire DNA sequences, or genomes of endangered species can be preserved, allowing the results of Nature's genetic experiment to be remembered *in silico*, and possibly reused in the future, even if that species is eventually lost.

Gene expression analysis

The expression of many genes can be determined by measuring mRNA levels with multiple techniques including microarrays, expressed cDNA sequence tag (EST) sequencing, serial analysis of gene expression (SAGE) tag sequencing, massively parallel signature sequencing (MPSS), or various applications of multiplexed in-situ hybridization. All of these techniques are extremely noise-prone and/or subject to bias in the biological measurement, and a major research area in computational biology involves developing statistical tools to separate signal from noise in high-throughput gene expression studies. Such studies are often used to determine the genes implicated in a disorder: one might compare microarray data from cancerous epithelial cells to data from non-cancerous cells to determine the transcripts that are up-regulated and down-regulated in a particular population of cancer cells.

Regulation analysis

Regulation is the complex orchestra of events starting with an extra-cellular signal and ultimately leading to the increase or decrease in the activity of one or more protein molecules. Bioinformatics techniques have been applied to explore various steps in this process. For example, promoter analysis involves the elucidation and study of sequence motifs in the genomic region surround the coding region of a gene. These motifs influence the extent to which that region is transcribed into mRNA. Expression data can be used to infer gene regulation: one might compare microarray data from a wide variety of states of an organism to form hypotheses about the genes involved in each state. In a single-cell organism, one might compare stages of the cell cycle, along with various stress conditions (heat shock, starvation, etc.). One can then apply clustering algorithms to that expression data to determine which genes are co-expressed. Further analysis could take a variety of directions: one 2004 study analyzed the promoter sequences of co-expressed (clustered together) genes to find common regulatory elements and used machine learning techniques to identify the promoter elements involved in regulating each cluster.

Protein expression analysis

Protein microarrays and high throughput (HT) mass

spectrometry (MS) can provide a snapshot of the proteins present in a biological sample. Bioinformatics is very much involved in making sense of protein microarray and HT MS data; the former approach faces similar problems as with microarrays targeted at mRNA, the latter involves the problem of matching large amounts of mass data against predicted masses from protein sequence databases, and the complicated statistical analysis of samples where multiple, but incomplete peptides from each protein are detected.

Analysis of mutations in cancer

Massive sequencing efforts are currently underway to identify point mutations in a variety of genes in cancer. The sheer volume of data produced requires automated systems to read sequence data, and to compare the sequencing results to the known sequence of the human genome, including known germline polymorphisms. Oligonucleotide microarrays, including comparative genomic hybridization and single nucleotide polymorphism arrays, able to probe simultaneously up to several hundred thousand sites throughout the genome are being used to identify chromosomal gains and losses in cancer. Hidden Markov model and change-point analysis methods are being developed to infer real copy number changes from often-noisy data. Further informatics approaches are being developed to understand the implications of lesions found to be recurrent across many tumors. Some modern tools (e.g. Quantum 3.1) provide tool for changing the protein sequence at specific sites through alterations to its amino acids and predict changes in the bioactivity after mutations.

Structure prediction

Protein structure prediction is another important application of bioinformatics. The amino acid sequence of a protein, the so-called *primary structure*, can be easily determined from the sequence on the gene that codes for it. In the vast majority of cases, these primary structures uniquely determine a structure in its native environment. (Of course, there are exceptions, such as the bovine spongiform encephalopathy - aka Mad Cow Disease - prion.) Knowledge of this structure is vital in understanding the function of the protein. For lack of better terms, structural information is usually classified as one of *secondary*, *tertiary* and *quaternary* structures. A viable general solution to such predictions remains an open problem. As of now, most efforts have been directed towards heuristics that work most of the time. One of the key ideas in bioinformatics research is the notion of homology. In the genomic branch of bioinformatics, homology is used to predict the function of a gene: if the sequence of gene *A*, whose function is known, is homologous to the sequence of gene *B*, whose function is unknown, one could infer that B may share A's function. In the structural branch of bioinformatics homology is used to determine which parts of the protein are important in structure formation and interaction with other proteins. In a technique called homology modeling, this information is used to predict the structure of a protein once the structure of a homologous protein is known. This currently remains the only way to predict protein structures reliably. One example of this is the similar protein homology between hemoglobin in humans and the hemoglobin in legumes (leg hemoglobin). Both serve the same purpose of transporting oxygen in both organisms. Though both of these proteins have completely different amino acid sequences, their protein structures are virtually identical, which reflects their near identical purposes. Other techniques for predicting protein structure include protein threading and *de novo* (from scratch) physics-based modeling.

Comparative genomics

The core of comparative genome analysis is the establishment of the correspondence between genes (orthology analysis) or other genomic features in different organisms. It is these inter-genomic maps that make it possible to trace the evolutionary processes responsible for the divergence of two genomes. A multitude of evolutionary events acting at various organizational levels shape genome evolution. At the lowest level, point mutations affect individual nucleotides. At a higher level, large chromosomal segments undergo duplication, lateral transfer, inversion, transposition, deletion and insertion. Ultimately, whole genomes are involved in processes of hybridization, polyploidization and endosymbiosis, often leading to rapid speciation. The complexity of genome evolution poses many exciting challenges to developers of mathematical models and algorithms, who have recourse to a spectra of algorithmic, statistical and mathematical techniques, ranging from exact, heuristics, fixed parameter and approximation algorithms for problems based on parsimony models to Markov Chain Monte Carlo algorithms for Bayesian analysis of problems based on probabilistic models. Many of these studies are based on the homology detection and protein families' computation.

Modeling biological systems

Systems biology involves the use of computer simulations of cellular subsystems (such as the networks of metabolites and enzymes which comprise metabolism, signal transduction pathways and gene regulatory networks) to both analyze and visualize the complex connections of these cellular processes. Artificial life or virtual evolution attempts to understand evolutionary processes via the computer simulation of simple (artificial) life forms.

High-throughput image analysis

Computational technologies are also used to accelerate or fully automate the processing, quantification and analysis of large amounts of high-information-content Biomedical imagery. Modern image analysis systems augment the observers' ability to make measurements from a large or complex set of images, by improving accuracy, objectivity, or speed. A fully developed analysis system may completely replace the observer. While these systems are not unique to biology related imagery, their application to biologic problems continue to provide unique challenges and solutions, placing several imagery application under the umbrella of Bioinformatics. These systems are in the process of becoming more important for both diagnostics and research. Some examples:

- high-throughput and high-fidelity quantification and sub-cellular localization (high-content screening, cytohistopathology)
- morphometrics are used to analyze pictures of embryos to track and to predict the fate of cell clusters during morphogenesis
- clinical image analysis and visualization
- determine the real-time air-flow patterns in breathing lungs of living individuals before and during challenge
- quantify occlusion size in real-time imagery from the development of and recovery during arterial injury
- making behavioral observations from extended video recordings of laboratory animals
- infrared measurements for metabolic activity determination

Software tools

The computational biology tool best-known among biologists is probably BLAST, an algorithm for searching large sequence (protein, DNA) databases. NCBI provides a popular implementation that searches their massive sequence databases. Bioinformatic meta search engines (Entrez, Bioinformatic Harvester) help finding relevant information from several databases. There are also free Web-based software designed for structural bioinformatics such as STING. Computer scripting languages such as Perl and Python are often used to interface with biological databases and parse output from bioinformatics programs. Communities of bioinformatics programmers have set up free/open source projects such as EMBOSS, Bioconductor, BioPerl, BioLinux, BioPython, BioRuby, and BioJava which develop and distribute shared programming tools and objects (as program modules) that make bioinformatics easier. An integrated software workbench consisting of many free/open source tools described above and many others is known as VigyaanCD. Taverna an open-source bioinformatics workbench that utilizes a workflow model of experimental design. Taverna is included as part of the myGRID package of e-science software. Quantum 3.1 is an example of the bioinformatics post-QSAR technology applying quantum and molecular physics instead of statistical methods. Genevestigator is an example of how large-scale gene expression microarray data is used to predict gene function based on contextual information. More recently, SOAP-based interfaces have been developed for a wide variety of bioinformatics applications such as blast, fasta, EMBOSS, clustalw, t-coffee, MUSCLE and many others. These are available from the EBI at EBI Web Services.

What is Biotechnology Science?

Biotechnology can be summarized as the manipulation

of living organisms to produce goods and services. Although the technology has received widespread media coverage in the past few years, it is a technology with a long history, dating as far back as 6000 B.C. Advances in science and technology have transformed traditional biotechnology techniques, such as selective breeding, hybridization and mutagenesis, into modern ones, such as recombinant DNA techniques and tissue culture. This transformation has opened the door to more varied applications in areas such as health care, the environment, forestry, industrial processes and others. Some developments to watch for include research into nutritionally enhanced GM foods and transgenic animals, bio chips and protein drugs.

Generally, any technique that is used to make or modify the products of living organisms in order to improve plants or animals, or to develop useful microorganisms. In modern terms, biotechnology has come to mean the use of cell and tissue culture, cell fusion, molecular biology, and in particular, recombinant deoxyribonucleic acid (DNA) technology to generate unique organisms with new traits or organisms that have the potential to produce specific products. Recombinant DNA technology has opened new horizons in the study of gene function and the regulation of gene action. In particular, the ability to insert genes and their controlling nucleic acid sequences into new recipient organisms allows for the manipulation of these genes in order to examine their activity in unique environments, away from the constraints posed in their normal host. Genetic transformation normally is achieved easily with microorganisms; new genetic material may be inserted into them, either into their chromosomes or into extra-chromosomal elements, the plasmids. Thus, bacteria and yeast can be created to metabolize specific products or to produce new products. Genetic engineering has allowed for significant advances in the understanding of the structure and mode of action of antibody molecules. Practical use of immunological techniques is pervasive in biotechnology. Few commercial products have been marketed for use in plant agriculture, but many have been tested. Interest has centered on producing plants that are resistant to specific herbicides. This resistance would allow crops to be sprayed with the particular herbicide, and only the weeds would be killed, not the genetically engineered crop species. Resistances to plant virus diseases have been induced in a number of crop species by transforming plants with portions of the viral genome, in particular the virus's coat protein. Biotechnology also holds great promise in the production of vaccines for use in maintaining the health of animals. Interferons are also being tested for their use in the management of specific diseases. Animals may be transformed to carry genes from other species including humans and are being used to produce valuable drugs. For example, goats are being used to produce tissue plasminogen activator, which has been effective in dissolving blood clots.

Plant scientists have been amazed at the ease with which plants can be transformed to enable them to express foreign genes. This field has developed very rapidly since the first transformation of a plant was reported in 1982, and a number of transformation procedures are available. Genetic engineering has enabled the large-scale production of proteins that have great potential for treatment of heart attacks. Many human gene products, produced with genetic engineering technology, are being investigated for their potential use as commercial drugs. Recombinant technology has been employed to produce vaccines from subunits of viruses, so that the use of either live or inactivated viruses as immunizing agents is avoided. Cloned genes and specific, defined nucleic acid sequences can be used as a means of diagnosing infectious diseases or in identifying individuals with the potential for genetic disease. The specific nucleic acids used as probes are normally tagged with radioisotopes, and the DNAs of candidate individuals are tested by hybridization to the labeled probe.

The technique has been used to detect latent viruses such as herpes, bacteria, mycoplasmas, and plasmodia, and to identify Huntington's disease, cystic fibrosis, and Duchenne muscular dystrophy. It is now also possible to put foreign genes into cells and to target them to specific regions of the recipient genome. This presents the possibility of developing specific therapies for hereditary diseases, exemplified by sickle-cell anemia. Modified microorganisms are being developed with abilities to degrade hazardous wastes. Genes have been identified that are involved in the pathway known to degrade polychlorinated biphenyls, and some have been cloned and inserted into selected bacteria to degrade this compound in contaminated soil and water. Other organisms are being sought to degrade phenols, petroleum products, and other chlorinated compounds.

Sub-fields of Biotechnology

Red biotechnology is biotechnology applied to medical processes. Some examples are the designing of organisms to produce antibiotics, and the engineering of genetic cures to cure diseases through genomic manipulation. White biotechnology, also known as grey biotechnology, is biotechnology applied to industrial processes. An example is the designing of an organism to produce a useful chemical. White biotechnology tends to consume less in resources than traditional processes when used to produce industrial goods. Green biotechnology is biotechnology applied to agricultural processes. An example is the designing of transgenic plants to grow under specific environmental conditions or in the presence (or absence) of certain agricultural chemicals. One hope is that green biotechnology might produce more environmentally friendly solutions than traditional industrial agriculture. An example of this is the engineering of a plant to express a pesticide, thereby eliminating the need for external application of pesticides. An example of this would be Bt corn. Whether or not green biotechnology products such as this are ultimately more environmentally friendly is a topic of considerable debate. Bioinformatics is an interdisciplinary field which addresses biological problems using computational techniques. The field is also often referred to as computational biology. It plays a key role in various areas like functional genomics, structural genomics, and proteomics amongst others, and forms a key component in biotechnology and pharmaceutical sector. The term blue biotechnology has also been used to describe the marine and aquatic applications of biotechnology, but its use is relatively rare.

Biotechnology can be understood as the manipulation of living organisms to produce goods and services. Although the technology has received widespread media coverage in the past few years, it is a technology with a long history, dating as far back as 6000 B.C.

The term biotechnology was coined in 1919 by a Hungarian engineer called Karl Ereky. However, its origins date back further than that. The history of biotechnology can be divided into two eras: traditional and modern biotechnology. Traditional biotechnology dates back thousands of years, to early farming societies in which people collected seeds of plants with the most desirable traits for planting the following year. This practice is now known as selective breeding. The same selective breeding practices were used by early Babylonians, Egyptians, and Romans to improve livestock. As far back as 6000 B.C., natural processes such as fermentation, in which microorganisms such as bacteria, yeasts and moulds play a critical role, were used to produce bread, beer and wine. Gregor Mendel's study of genetics, using seed and plant experiments at the end of the 19th century, gave the first indications of the cross from traditional to modern biotechnology. He discovered that traits are transmitted from parents to offspring by discrete, independent units, later called genes. His observations laid the groundwork for the field of genetics.

In 1943, the first direct evidence that deoxyribonucleic acid, or DNA, carried genetic information was discovered. However, it wasn't until 1953 that the mystery of the structure of DNA and the way genetic information is passed from generation to generation was unlocked by the discovery of the structure of DNA by James Watson and Francis Crick. The era of 'modern biotechnology', which involves manipulation of genes from living organisms in more precise and controlled ways than traditional biotechnology, began with their discovery. In 1985, genetically engineered plants resistant to insects, viruses, and bacteria were tested for the first time. Since then, many genetically engineered plants have been developed, successfully field tested and received food, livestock feed and environmental safety approval in many countries, including Canada. It was also in 1985 that a plan for mapping and sequencing the human genome was made. The goal of this Human Genome Initiative, which was launched in 1990, was to map all of the 80,000 to100,000 human genes by the year 2003.

Select Key Visionaries and Personalities in Biotechnology Sector

Leena Palotie

Professor M.D. Leena Palotie (born June 16 1952 in Helsinki, Finland) is a pioneering Finnish researcher of genetics. Palotie is considered to be one of the leading researchers in the field of genetics. She has among others participated in discovering 15 genes in the genotype of Finnish inherited diseases, such as arterial hypertension, schizophrenia, lactose intolerance, arthrosis and multiple sclerosis. From 1998 to 2002 Palotie participated in founding the UCLA Human Genetics Genome Center and has worked as a professor in the Academy of Finland since 2003. In April of 2005 Palotie was employed in the University of Helsinki and the National Public Health Institute of Finland. She is also project director in the EU project GenomEUtwin that was formed to define and characterize the genetic components in the background of different diseases. Since 2004, she has been a member of the Board of Directors of Orion Corporation, the largest Finnish pharmaceutical company. Palotie has published over 370 research articles and around 60 invite articles. In addition to many academic rewards she has received a honorary degree from the University of Uppsala.

Kari Stefansson

Dr. Kári Stefánsson, M.D., Dr.Med. from the University of Iceland, is the Chairman, CEO and co-founder of deCODE Genetics and a former professor of neurology, neuropathology and neuroscience at Harvard University (1993-1997). From 1993-1996 he was director of neuropathology at Boston's Beth Israel Hospital. Dr. Kári also held faculty positions at the University of Chicago. Dr. Kári opened the NASDAQ Stock Market on July 20, 2005, because of deCode's five years anniversary on the NASDAQ Market.

Craig Venter

Venter, Craig [b. Salt Lake City, Utah, October 14, 1946] developed a method of deciphering genomes known as whole-genome shotgun sequencing. The genome to be analyzed is broken into random, overlapping fragments of

DNA that are a few thousand letters in length. Each fragment is sequenced, or read. Then the fragments are reassembled by a computer into their correct order. Although there were initially many skeptics, Venter's conviction that shotgunning would be faster and just as accurate for much genome deciphering proved to be true, and the technique is now widely used. In 1995 Venter and his team used the technique to obtain the first complete genome (DNA sequence) of an organism other than a virus, that of the bacterium *Haemophilus influenzae.* In 2000, in collaboration with researchers at the University of California at Berkeley, he published almost the entire genome of the fruit fly *Drosophila melanogaster.* In 2001 his team and a competing group published rough drafts of the human genome.

Sydney Brenner

Sydney Brenner, CH (born January 13, 1927) is a British biologist active in the United States. Born in Germiston, South Africa, he made seminal contributions to the emerging field of molecular biology in the 1960s, notably in the elucidation of the triplet code of protein translation through the Crick, Brenner et al. experiment of 1961, which discovered frameshift mutations. This insight provided early elucidation of the genetic code. Brenner then turned his sights on establishing *Caenorhabditis elegans* as a model organism for the investigation of animal development including neural development. Brenner chose this 1 millimeter-long soil roundworm mainly because it is simple, is easy to grow in bulk populations, and turned out be quite convenient for genetic analysis. The title of his Nobel lecture on December 2002, "Nature's Gift to Science" is an homage to this modest nematode, and he considered that having chosen the right organism turned out to be as important as having addressed the right problems to work on. For the latter work he shared the 2002 Nobel Prize in Physiology or

Medicine with H. Robert Horvitz and John Sulston. Brenner founded the Molecular Sciences Institute and is currently associated with the Salk Institute and the Institute of Molecular and Cell Biology. Known for his penetrating scientific insight and acerbic wit, Brenner for many years penned a regular column ("Loose Ends") in the journal *Current Biology*; he wrote "A Life In Science" (ISBN 0954027809) paperback published by Biomed Central Ltd. in 2001:

Eric Lander

Eric Steven Lander (b. February 3, 1957) is a Professor of Biology at the Massachusetts Institute of Technology (MIT), a member of the Whitehead Institute, and director of the Broad Institute of MIT and Harvard who has devoted his career toward realizing the promise of the human genome for medicine. He graduated from Stuyvesant High School in 1974 and then attended Princeton University. At the age of seventeen, he wrote a paper on quasiperfect numbers for which he won the Westinghouse Prize. He wrote his doctorate on symmetric designs at Oxford University as a Rhodes Scholar. As a Mathematician he studied combinatorics and applications of group representational theory to coding theory. He enjoyed mathematics but did not wish to spend his life in such a "monastic career." Unsure of what to do next, he took up a job teaching managerial economics at Harvard Business School; he also began to write a book on information theory. At the suggestion of his brother, Arthur Lander, he started to look at neurobiology "because there's a lot of information in the brain". In order to understand mathematical neurobiology, he felt he had to study cellular neurobiology; this in turn led to studying microbiology and continued down to the level of genetics. "When I finally feel I have learned genetics, I should get back to these other problems. But I'm still trying to get the genetics right". His studies introduced him to David Botstein, a geneticist working at MIT. Botstein was working on a way to unravel how subtle differences in complex genetic systems can become disorders like cancer, diabetes, schizophrenia, and even obesity. Lander then joined Whitehead Institute (1986) and later joined MIT as a geneticist. In 1990 he founded W.I.C.G.R. (Whitehead Institute/MIT Center for Genome Research) W.I.C.G.R. was one of the world's leading centers of genome research, and under Doctor Lander's leadership, it has made great progress in developing new methods of analysing mammalian genomes. The Whitehead institute has also made important breakthroughs in applying this information to the study of human variation and particularly the study of medical genetics. The W.I.C.G.R. formed the basis for the foundation of the Broad Institute, a transformation in which Dr. Lander was instrumental. There were two main groups attempting to sequence the human genome: the first was the Human Genome Project, the public funded effort which intended to publish the information obtained freely open for all to use without restrictions. This was a collaborative effort involving many research groups from countries all over the world. The second effort was undertaken by Celera Genomics who intended to patent the information obtained and charge subscriptions for use of the sequence data (Celera has since abandoned this policy and has donated large amounts of sequence information for free public use) The H.G.P. was established first but moved slowly and when Celera entered the race they were able to take the lead. In the end the H.G.P. team published first and much of the credit goes to the W.I.C.G.R. They led the effort to develop genetic and physical maps of human and mouse genomes. This was an essential step to the clone by clone method of sequencing used in the H.G.P. The W.I.C.G.R. team sequenced approximately one third of the Human Genome making them the primary contributors to the H.G.P. This earned Lander the honour of being named first author when the draft of the human genome was published in 2001. For his work he was named one of Time magazine's 100 most influential people of our time (2004). Lander has also appeared in numerous PBS documentaries about genetics.

The WICGR has also made a leading contribution to the sequencing of the mouse genome. Aside from academic interest this is an important step in fully understanding the molecular biology of mice which are often used as model organisms in studies of everything from human diseases to embryonic development. Increased understanding of mice will thus facilitate many areas of research. The W.I.G.C.R. has also sequenced the genomes of Ciona intestinalis, the pufferfish, the filamentous fungus Neurospora crassa and multiple relatives of Saccharomyces cerevisiae one of the most studied yeasts. The Ciona intestinalis genome provides a good system for exploring the evolutionary origins of all vertebrates. Pufferfish have smaller sized genomes compared to other vertebrates; as a result their genomes are "mini" models for vertebrates.

The sequencing of the yeasts related to Saccharomyces cerevisiae will ease the identification of key gene regulatory elements some of which may be common to all eukaryotes (including both plant and animal kingdoms). Sequence data is just that a list of bases found in a given stretch of DNA, its value lies in the application of this information in the discoveries and new technologies it allows. In Dr. Lander's case the application was the study of disease he is the founder director of the Broad Institute this is a collaboration between MIT, Harvard, the Whitehead institute and affiliated hospitals its goal is "to create tools for genome medicine and make them broadly available to the scientific community; to apply these tools to propel the understanding and treatment of disease". To this end they are studying the variation in the human genome and have led an international effort which ha assembled a library of 2.1 million Single nucleotide polymorphisms (SNP) these act as markers or signposts in the genome allowing the identification of disease susceptibility genes. They hope to construct a map of the human genome using blocks of these SNP called Linkage disequilibrium or LD this map will be of significant help in medical genetics. It will allow researchers to link a given condition to a given gene or set of genes using the LD as a marker this will allow for improved diagnostic procedures. Lander and his colleagues are hoping the LD map will allow them to test the Common Disease-Common Variant hypothesis which states than many common diseases may be caused by a small number of common alleles, for example 50% of the variance in susceptibility to Alzheimer's disease is explained by the common allele ApoE4. Landers group have recently discovered an important association which accounts for a large proportion of population risk for adult onset diabetes.

They are also using family studies to identify the genes involved in many genetic diseases like inflammatory bowel disease. Lander's most important work may be his development of a molecular taxonomy for cancers. The cancers are grouped according to gene expression and information like their response to chemotherapy is collected for each group. The division of cancers into homogeneous subgroups will allow increased understanding of the molecular origins of these cancers and aid the design of more effective therapies. They have also identified a new type of leukaemia called MLL and have identified a gene which may serve as a target for a new drug. Much of Lander's work is in genomics, a field which involves huge groups of researchers collaborating on large scale experiments. It is often difficult to single out one person as the driving force behind key discoveries, but Lander is an exception. He founded and directed WICGR, one of the worlds leading centres for genome research. His contribution to the Human Genome Project alone makes Lander's career noteworthy. He has continues to push the forefront of applying genomic knowledge to the difficult challenges of medical genetics, to which end he founded the Broad Institute. In addition to his research, he has for several years co-taught MIT's required undergraduate introductory biology course (7.012) with Robert Weinberg.

Leroy Hood

Leroy Hood is an American biologist. He won the 2003 Lemelson-MIT Prize for inventing "four instruments that have unlocked much of the mystery of human biology" by helping decode the genome. Hood also won the 2002 Kyoto Prize for Advanced Technology, and the 1987 Albert Lasker Award for Basic Medical Research. His inventions include the automated DNA sequencer, a device to create proteins and an automated tool for synthesizing DNA. Hood co-founded the Institute for Systems Biology. Dr. Leroy Hood was born October 10, 1938 in Missoula, Montana. He is recognized as one of the world's leading scientists in molecular biotechnology and genomics. He holds numerous patents and awards for his scientific breakthroughs and prides himself on his life-long commitment to making science accessible and understandable to the general public, especially children. One of this foremost goals is bringing hands-on, inquiry-based science to K-12 classrooms. Dr. Hood earned an M.D. From Johns Hopkins University in 1964 and a Ph.D. in biochemistry from the California Institute of Technology in 1968. Since then, his research has focused on the study of molecular immunology and biotechnology. Dr. Hood has published more than 500 peer-reviewed papers, received 12 patents, and co-authored textbooks in biochemistry, immunology, molecular biology, and genetics, and is a member of the National Academy of Sciences, the American Philosophical Society and the American Association of Arts and Sciences. Hood received a D.Sc. from Bates College in 1999. His professional career began at Caltech where he and his colleagues pioneered four instruments—the DNA gene sequence and synthesizer, and the protein synthesizer and sequencer—which comprise the technological foundation for contemporary molecular biology. In particular, the DNA sequencer has revolutionized genomics by allowing the rapid automated sequencing of DNA. Dr. Hood was also one of the first advocates of and is a key player in the Human Genome Project—the quest to decipher the sequence of the human DNA. He also played a pioneering role in deciphering the secrets of antibody diversity. In 1992, Dr. Hood moved to the University of Washington to create the cross-disciplinary Department of Molecular Biotechnology. In his role as the William Gates, III Professor of Biomedical Science, Dr. Hood applied his laboratory expertise in DNA sequencing to the analysis of human and mouse immune receptors and initiated studies in prostate cancer, autoimmunity, and hematopoietic stem cell development. In 2000, Dr. Hood co-founded the Institute for Systems Biology in Seattle, Washington to pioneer systems approaches to biology and medicine. He serves as President of the Institute and continues to pursue his interest in biology, medicine, technology, development, and computational biology. Dr. Hood has played a role in founding numerous biotechnology companies, including Amgen, Applied Biosystems, Systemix, Darwin, Rosetta, and MacroGenics.

Robert Langer

Robert S. Langer (born August 29, 1948 in Albany, New York) is an Institute Professor at the Massachusetts Institute of Technology. He was formerly the Germeshausen

Professor of Chemical and Biomedical Engineering and maintains activity in the departments of chemical engineering and the biological engineering division at MIT. He is a distinguished and highly regarded researcher in biotechnology, especially in the fields of drug delivery systems and tissue engineering. Dr. Langer's research laboratory at MIT is the largest biomedical engineering lab in the world, maintaining about $6 million in annual grants and over 100 researchers. Langer's contributions to medicine and the emerging fields of biotechnology are highly recognized and respected around the world. He is considered a pioneer of many new technologies, including transdermal delivery systems, which allow the administration of drugs or extraction of analytes from the body through the skin without needles or other invasive methods. He and the researchers in his lab have also made significant advances in tissue engineering, such as the creation of vascularized engineered muscle tissue and engineered blood vessels. Langer holds more than 500 granted or pending patents and has authored more than 800 scientific papers. He has proven adroit at bringing together academia and industry and has participated in the founding of more than two dozen companies. He has received numerous awards, including the Charles Stark Draper Prize, the Lemelson-MIT Prize and the Albany Medical Center Prize in Medicine and Biomedical Research. He is the youngest person in history (at 43) to be elected to all three American science academies: the National Academy of Sciences, the National Academy of Engineering and the Institute of Medicine. Dr. Langer received his bachelor's degree from Cornell University in chemical engineering. He earned his Sc.D. in chemical engineering from MIT in 1974. His dissertation was entitled "Enzymatic regeneration of ATP" and his advisors were Clark K. Colton and Michael Archer. From 1974-1977 he worked as a postdoctoral fellow for famed cancer researcher Judah Folkman at the Children's

Hospital Boston and at Harvard Medical School. He has been an advisor to influentical scholars in the fields of biomaterials, drug delivery systems, and tissue engineering: W. Mark Saltzman, David J. Mooney, and Lisa E. Freed.

Herbert Boyer

Herbert (Herb) W. Boyer (born 1936) is a co-recipient of the 1996 Lemelson-MIT Prize and a co-founder of Genentech. Boyer received his bachelor's degree in biology and chemistry from St. Vincent's College in Latrobe, Pennsylvania in 1958. He married his wife Grace the following year. He received his PhD at the University of Pittsburgh in 1963 and participated as an activist in the civil rights movement. He spent three years in post-graduate work at Yale University, then became an assistant professor at the University of California, San Francisco, where he discovered that bacteria could be combined with genes from higher organisms. In August 1978, he produced the first synthetic insulin using his new transgenic bacteria, followed in 1979 by a growth hormone. Genentech's approach to the first synthesis of insulin won out over Wally Gilbert's approach at Biogen which used genes from natural sources. Boyer created his gene de novo from its individual nucleotides. Boyer's discoveries and collaboration with Robert Swanson to found Genentech is featured in the history of innovation *They Made America* by Harold Evans (Little Brown, 2004) and in the subsequent WGBH television series.

Michael West

Dr. Michael D. West is the Chairman of the Board and the Chief Scientific Officer of Advanced Cell Technology Corporation, which specializes in Stem Cell research. Dr. West was the founder of Geron Corporation and served as its Director and Senior Executive Officer from 1990 to 1998. Dr. West's 2003 book The Immortal Cell (ISBN 0385509286) tells his personal story of his religious, business and scientific struggles to deal with the problem of human aging.

Thomas Okarma

Dr. Thomas Okarma is the current CEO of Geron Corporation, a biotechnology company. Dr. Okarma earned his undergraduate degree from Dartmouth College and his Master's & Ph.D. from Stanford University. Dr. Okarma worked at the Stanford University School of Medicine as a faculty member from 1980 until 1985. He left the university to become the scientific founder of Applied Immune Sciences. He worked as the Vice President of Reserach and Development and later as the CEO until the company was acquired by Rhone-Poulenc Rorer. Okarma joined Geron Corporation in 1997, serving initially as Vice President of Cell Therapies, and later as Vice President of Research and Development. When founder and CEO Dr. Michael West left the to company to start Advanced Cell Technology in 1999 Dr. Okarma was promoted to CEO, a position that he continues to hold seven years later.

Kiran Mazumdar-Shaw (Biocon)

Dr. Kiran Mazumdar-Shaw (b. 23 March, 1953 in Bangalore) is an Indian entrepreneur. She is the Chairman & Managing Director of Biocon Ltd. In 2004, she became India's richest woman. She was educated at the Bishop Cotton Girls School and Mount Carmel College at Bangalore. After obtaining a B.Sc. Honours degree in Zoology from Bangalore University in 1973, she joined the Ballarat University in Melbourne, Australia and qualified as a master brewer in 1975 to become India's first woman Brew master. Her father encouraged her in this profession, as he himself was a master brewer in India. Her professional career started with the position of trainee brewer in Carlton & United Beverages in 1974. During 1975-77, she worked in technical positions in Kolkata and Vadodara. In 1978, she joined as Trainee Manager with Biocon Biochemicals Limited in

Ireland. Collaborating with the same Irish firm, she founded Biocon India with a capital of Rs. 10,000/- in her garage in 1978. The initial operation was to extract an enzyme from papaya. Her application for loans was turned down by banks on three counts – biotechnology was a new word yet; the company lacked assets and most importantly, women entrepreneurs were still a rarity. On account of the last reason, she faced problems in recruiting as well. Over the years, the company grew under her stewardship and is today the biggest biopharmaceutical firm in India. In 2004, Biocon went for an IPO and the issue was over-subscribed by over 30 times. Post-IPO, Shaw held close to 40% of the stock of the company and was regarded as India's richest woman with an estimated worth of Rs. 2,100 crore (~U.S. $ 480 million). She is a civic activist, especially with respect to municipal administration in Bangalore. She is also an art collector. She has authored 'Ale and Arty,' a Coffee table book about brewing beer illustrated by paintings of some of India's renowned artists. Famous brewing families and beer firms are the subject of the book. In 1998, she married John Shaw, an expatriate manager and Indophile from Scotland at Madura Coats. John Shaw resigned as the managing director of Madura Coats the same year and joined Biocon as its Director for International Business and the Vice Chairman of the Board.

Biotechnology Applications

Modern biotechnology techniques are currently being used in many areas such as food, agriculture, health care, forestry, the environment, minerals, oil and gas recovery and industrial processes to develop new products and processes, and to modify existing ones. Summarized below are some of the ways that modern biotechnology techniques have been applied in these areas.

Food and Agriculture

One of the most extensive applications of biotechnology has been in agriculture. Biotechnology techniques such as recombinant DNA techniques and mutagenesis have been used to develop plants with novel traits. These traits include herbicide tolerance and pest, insect and virus resistance. Biotechnology techniques have been used to produce biopesticides which are toxic to targeted plant pests. Also, experiments into genetic modification of aquatic organisms such as salmon, for such novel traits as enhanced growth, have been carried out using recombinant DNA techniques.

Health Care

To date, applications of biotechnology in health care have focused on fighting disease using the human body's own 'weapons'. Biotechnology medicines and therapies synthesize proteins, enzymes, antibodies and other substances which occur naturally in the human body, to fight infections and diseases. However, biotechnology also uses other living organisms, that is, plant and animal cells, viruses and yeasts to help produce human medicines. There are four main areas in health care in which biotechnology is currently being used: medicines, vaccines, diagnostics and gene therapy. Several biotechnology medicines have been developed and approved to treat such diseases as anaemia, growth deficiency in children, hemophilia, diabetes and others. Biotechnology companies have also developed therapeutic products for infectious agents such as HIV, hepatitis B and influenza. Biotechnology is responsible for new vaccines such as the hepatitis B vaccine which was approved in the U.S. to fight the hepatitis B virus. Biotechnology diagnostics have been used to detect a wide variety of diseases and genetic conditions. For example, gene therapy is used diagnostically to screen donated blood to protect the blood supply from HIV and hepatitis. Biotechnology diagnostics are also used in home pregnancy tests. In a home pregnancy test kit, a protein called a monoclonal antibody (MAb), binds to Human Chorionic Gonadotrophin (HCG), causing a colour change. HCG is present in a woman's urine only during pregnancy. In gene therapy, genes are used as 'drugs' to treat hereditary genetic disorders. Faulty or missing genes can be replaced to prevent the occurrence of a genetic disease. Current gene therapy is primarily experiment-based, with a few human clinical trials, such as for the treatment of cystic fibrosis, in the early stages.

Environment

Biotechnology applications in the environment focus on using living organisms primarily to treat waste and prevent pollution. Examples of these applications include bio-filtration and bio-remediation. Bio-filtration refers to the use of microorganisms to remove complex pollutants from the air emissions and waste water discharges of many manufacturing processes. Bio-remediation refers to a number of processes that use living organisms to degrade toxic waste into harmless by-products such as water, carbon dioxide and other materials. Examples of bio- remediation processes include:

- *Bio-stimulation:* A technique which involves introducing nutrients to stimulate the growth of waste-eating microorganisms already present in the environment at a waste site.
- *Bio-augmentation:* A technique whereby microorganisms, not normally present in the ecosystem of a contaminated site, are added to the site to clean up pollutants. The microorganisms added may be natural or genetically modified. The organisms die when their food is used up.
- *Phytoremediation:* A technique which uses certain

plants and fungi, planted at waste sites, to naturally reclaim the sites without using chemical treatment, incineration or land filling. These plants and fungi are used because they flourish by accumulating metals and other waste materials present in the soil.

Forestry

To date, only a few biotechnology-derived products and processes used in forestry have been commercialized. Most biotechnology applications in forestry are still in the research and development stage. However, some commercial applications include the development of bacterial bio-pesticides such as *Bacillus thuringiensis* (Bt), as an alternative to chemical pesticides. Tissue culture technology is also being used to produce genetically modified seedlings for forest regeneration. For example, BCRI Inc., a biotechnology company in British Columbia, is using tissue culture to produce weevil-resistant Sitka spruce. Companies are also developing bio-fertilizers such as plant growth-promoting rhizobia bacterial fertilizers. In addition, genetic engineering has been used to develop traits of agronomic interest such as pest and disease resistance, within trees. The pulp and paper industry has used enzymes and microorganisms in their bio-bleaching and de-inking processes.

Industrial Processes

Biotechnology has been applied in a variety of industrial processes in different ways, particularly in the use of biocatalysts in manufacturing processes. Biocatalysts are substances that initiate or modify the rate of a biological process and are generally consumed in the process. Some examples of industrial processes where biotechnology has been applied include chemicals, starch/grain processing, cleaning and textile industrial processes. In the chemicals industry, biotechnology has been used to produce commodity and specialty chemicals. It is also used in the starch and grain processing industries through the use of enzymes to turn starch into glucose and fructose. Corn and other grains can be converted to sweeteners such as high-fructose corn syrup and maltose syrup, using enzymes. Biotechnology applications have been used to produce ethanol from grain. Furthermore, biotechnology is used in the textile industry for the finishing of fabrics and garments. In the pulp and paper industry, bio-pulping is used to manufacture some products.

Metals and Minerals Recovery

Biotechnology has been used in the mining industry in such processes as bio-leaching and metal bio-remediation and recovery. Bio-leaching is the use of bacteria to extract valuable metals. Metal bio-remediation and recovery refers to the use of biologically created enzymes in the degreasing process to recover metals of value.

Energy

To date, biotechnology applications in the energy industry include the production of cleaner coal and petroleum by removing sulfur and thus reducing the environmental contaminants released during combustion. Biotechnology has also been used widely to produce bio fuels such as bio-ethanol and bio-diesel. Bio fuels are alcohols, ethers, esters and other organic chemicals made from biomass such as herbaceous and woody plants, agricultural and forestry residues and industrial waste. These fuels can be used for electricity and as fuels for transportation.

Human Applications

When treating problems that arise from genetic disorder, one solution is gene therapy. A genetic disorder is a situation where some genes are missing or faulty. When this happens, genes may be expressed in unfavorable ways or not at all, and this generally leads to further complications. The idea of gene therapy is that a non-pathogenic virus or other delivery system can be used to insert a piece of DNA—a good copy of the gene—into cells of the living individual. The modified cells would divide as normal and each division would produce cells that express the desired trait. The result would be that he/she would then have the ability to express the trait that was previously absent at least partially. This form of genetic engineering could help alleviate many problems, such as diabetes, cystic fibrosis, or other genetic diseases.

The potential of genetic engineering to cure medical conditions opens the question of exactly what such a condition is. Some advocates see aging and death as medical conditions and engineering problems to be solved. They see human genetic engineering as a key tool in this *(see life extension)*. The difference between cure and enhancement from their perspective is merely one of degree. Theoretically genetic engineering could be used to drastically change people's genomes which could enable people to regrow limbs, the spine, the brain. It could also be used to make people stronger, faster, smarter, or to increase the capacity of the lungs, among other things. If a gene exists in nature, it could be brought over to a human cell. Others feel that there is an important distinction between using genetic technologies to treat those who are suffering and to make those who are already healthy superior to others. There is widespread agreement that germline engineering should not currently be allowed for either therapeutic and enhancement applications, as evidenced by a recent report by the American Association for the Advancement of Science. Though theory and speculation suggest that genetic engineering could be used to make people stronger, faster, smarter, or to increase lung capacity, the AAAS report finds that there is little evidence that this currently can be done safely or without unethical human experiments.

"Transgenic organism" is now the preferred term for genetically modified organisms with extra-genome (foreign genetic) information, as opposed to "genetically engineered" or "genetically modified" organisms (which may refer to changes made within the genome such as amplification or deletion of genes). The first Genetically Engineered drug was human insulin approved by the USA's FDA in 1982. Another early application of GE was to create human growth hormone as replacement for a drug that was previously extracted from human cadavers. In 1986 the FDA approved the first genetically engineered vaccine for humans, for hepatitis B. Since these early uses of the technology in medicine the use of the GE has expanded to supply many drugs and vaccines. One of the best known applications of genetic engineering is that of the creation of genetically modified organisms (GMOs). There are potentially momentous biotechnological applications of GM, for example oral vaccines produced naturally in fruit, at very low cost. A radical ambition of some groups is human enhancement via genetics, eventually by molecular engineering. DNA sequencing is a technique which is used to identify each base in DNA. Although the costs of DNA sequencing has dropped dramatically, the NIH estimates it costs at least $10 million to sequence 3 billion base pairs - the size of the whole human genome. Although there has been a tremendous revolution in the biological sciences in the past twenty years, there is still a great deal that remains to be discovered. The completion of the sequencing of the human genome, as well as the genomes of most agriculturally and scientifically important plants and animals, has increased the possibilities of genetic research immeasurably. Expedient and inexpensive access to comprehensive genetic data has become a reality, with billions of sequenced nucleotides already online and annotated. Now that the rapid sequencing of arbitrarily large genomes has become a simple, if not trivial affair, a much greater challenge will be elucidating function of the extraordinarily complex web of interacting proteins, dubbed the proteome, that constitutes and powers all living things. Genetic engineering has become the gold standard in protein research, and major research progress has been made using a wide variety of techniques, including:

- Loss of function, such as in a knockout experiment, in which an organism is engineered to lack the activity of one or more genes. This allows the experimenter to analyze the defects caused by this mutation, and can be considerably useful in unearthing the function of a gene. It is used especially frequently in developmental biology. A knockout experiment involves the creation and manipulation of a DNA construct in vitro, which, in a simple knockout, consists of a copy of the desired gene which has been slightly altered such as to cripple its function. The construct is then taken up by embryonic stem cells, where the engineered copy of the gene replaces the organism's own gene. These stem cells are injected into blastocysts, which are implanted into surrogate mothers. Another method, useful in organisms such as Drosophila (fruit fly), is to induce mutations in a large population and then screen the progeny for the desired mutation. A similar process can be used in both plants and prokaryotes.
- Gain of function experiments, the logical counterpart of knockouts. These are sometimes performed in conjunction with knockout experiments to more finely

establish the function of the desired gene. The process is much the same as that in knockout engineering, except that the construct is designed to increase the function of the gene, usually by providing extra copies of the gene or attracting more frequent transcription.

- 'Tracking' experiments, which seek to gain information about the localization and interaction of the desired protein. One way to do this is to replace the wild-type gene with a 'fusion' gene, which is a juxtaposition of the wild-type gene with a reporting element such as Green Fluorescent Protein (GFP) that will allow easy visualization of the products of the genetic modification. While this is a useful technique, the manipulation can destroy the function of the gene, creating secondary effects and possibly calling into question the results of the experiment. More sophisticated techniques are now in development that can track protein products without mitigating their function, such as the addition of small sequences which will serve as binding motifs to monoclonal antibodies.

Industrial biotechnology (also known as white biotechnology) is the practice of using cells to generate industrially-useful products. *The Economist* speculated (as cited in the *Economist* article listed in the "References" section) industrial biotechnology might significantly impact the chemical industry. *The Economist* also suggested it might enable economies to become less dependent on fossil fuels. Diversa is an example of a company that specializes in industrial biotechnology. A significant problem in industrial biotechnology is waste production. A cell may be used to generate desirable carbon dioxide, other cells, and other molecules. It will use energy to accomplish its industrial purpose. Yet it will also use some energy to generate waste (like acetic acid) instead of the desired product or products. Decreasing waste production is a significant goal in industrial biotechnology. Metabolic engineering may help reach that goal.

Related Technologies and Biotech Industries

This looks at how biotechnology contributes to new and enhanced industrial processes. It examines how biotechnology enables the use of renewable resources in industry, thereby increasing efficiency, decreasing pollution, and reducing energy usage and waste production during various types of industrial processing.

- Biochemicals
- Bioplastics and Biopolymers
- Biotechnology in the Textiles Industry
- Enzymes in Industrial Applications
- Microbial Enhanced Oil Recovery
- Starch Bioprocessing

(a) Biochemicals

Biochemicals are chemicals produced from organic sources known as biomass. They are sometimes called bio-based chemicals, green chemicals, and plant-based chemicals. Biochemicals are chemicals produced from organic sources known as biomass. Biomass is any living matter that can be naturally and regularly replenished, including agricultural food and feed crop residues, aquatic plants, animal wastes and other waste materials. Biochemicals are sometimes called bio-based chemicals, green chemicals, and plant-based chemicals. Currently, the most popular source for biochemicals is corn. Chemicals like ethyl lactate, succinic acid, and polylactic acid are made from corn components, such as corn starch, and are used to make a variety of products, such as plastics, solvents, clothing fibres, paints, and food additives.

Petroleum is the oil found underground in certain parts of the world and is the most used energy source in the world. It is used to make gasoline, chemicals, plastics and drugs. Because petroleum is limited in quantity, and cannot be made once the supply has been depleted from the earth, it is said to be non-renewable. Petroleum products are also not environmentally friendly – causing air pollution when burned, and filling up landfills when disposed of. Biochemical products made from biomass help solve many of the problems associated with petrochemical products. Environmental benefits include less pollution from crude oil extraction and processing, as well as better waste management since the materials are biodegradable. And because biomass can be replenished regularly by nature, biochemical products are said to be renewable.

- *ommodity chemicals — These include solvents, fuel additives, lubricants, surfactants, adhesives, and inks.*
- *Fine chemicals — These include enzymes, nutraceuticals, and pharmaceuticals.*
- *Chemical intermediates — These include sugars, organic acids, and other monomers.*

Biochemicals are made from the processing of plant sugars. Currently, corn is being used extensively as a starting material for biochemicals. Making biochemicals involves two basic steps:

- First, organic matter, such as corn, is milled and the sugar is extracted from the raw material. The sugar is then fermented in a controlled environment, much like the fermentation of beer and wine. Some fermentation processes use genetically modified organisms that convert corn-derived sugar to a specific chemical.

- Next, a purification process is undertaken where the fermented liquid is purified for the target biochemical through methods such as distillation and electrodialysis. The biochemicals are then made into various products, such as plastics and solvents.

The role of biotechnology in the creation of biochemicals is crucial. Through fermentation technology, biotechnology enables the processing of common crops and crop wastes, such as corn kernels and corn stalks, to make products such as plastics, textiles, and ethanol. Biotechnology also allows the creation of genetically modified bacteria that can convert plant sugars into an ingredient used in clothing, packaging, and plastics. New enzymes are also being developed for breaking down plant sugars. Biotechnology also helps scientists better understand the biology of bacteria and enzymes that can be used in bioprocessing, which may enable the development of cheaper methods of biochemical production. Currently, biochemical products are often expensive to produce and are not cost-effective enough to compete with traditional products.

esearch in biochemicals continues to grow rapidly. Many large multinational companies are investing time and money into producing chemicals that can be economically viable and environmentally friendly. For example, many chemical companies are teaming up with biotechnology companies to develop products based on biochemicals. Cargill Dow currently makes polylactic acid from corn kernels but has plans to switch to cheaper feedstocks, such as corn stalks, wheat straw, rice hulls, and sawdust, so that biochemicals can be more competitive in the marketplace. Current research is also focussed on developing genetically modified microorganisms for use in specific chemical productions. Pollution due to the use of petroleum-based products may be eliminated by the use of biochemicals. Whereas petroleum-based products do not biodegrade and accumulate in waste landfills, biochemicals degrade completely and do not contribute to waste problems. As well, biochemicals are made from biomass; therefore, their production and use are sustainable.

(b) Biopolymers and Bioplastics

Many everyday products used by Canadians are made of plastic. Pens, computers, cars, food packaging, and clothing are all examples of products which contain plastic. Plastic is a material made up of one or more polymers. The main source of the chemicals needed to manufacture plastics are fossil fuels. These petrochemical plastics are very durable, but take a long time to biodegrade when disposed. Rising concern about the cost of fossil fuels, and their impact on the environment has resulted in a search for alternatives to petrochemical plastics, namely biopolymers and bioplastics.

Biopolymers and bioplastics go by many different names. They are often referred to as bio-based plastics and polymers, or as biodegradable plastics or polymers. They are defined below:

- Biopolymers are polymers which are present in, or created by, living organisms. These include polymers from renewable resources that can be polymerized to create bioplastics.
- Bioplastics are plastics manufactured using biopolymers, and are biodegradable.

Biopolymers and bioplastics are not new products. Henry Ford developed a method of manufacturing plastic car parts from soybeans in the mid-1900s. However, World War II side-tracked the production of bioplastic cars. Today, bioplastics are gaining popularity once again as new manufacturing techniques developed through biotechnology are being applied to their production.

There are two main types of biopolymers: those that come from living organisms; and, those which need to be polymerized but come from renewable resources. Both types are used in the production of bioplastics. These biopolymers are present in, or created by, living organisms. These include carbohydrates and proteins. These can be used in the production of plastic for commercial purposes.

(c) Biotechnology in the Textiles Industry

Biotechnology has impacted the textiles industry through the development of more efficient and more environmentally friendly manufacturing processes, as well as through the design of improved textile materials. Some of biotechnology's key roles have involved the implementation, production, and modification of enzymes for the improvement of textile manufacturing processes. Biotechnology has also facilitated the production of novel and biodegradeable fibres from biomass feedstocks.

Enzyme Biotechnology in Textiles

Through biotechnology, enzymes are used to treat and modify fibres during textile manufacturing, processing, and in caring for the product afterwards. Some applications include:

- De-sizing of cotton – Untreated cotton threads can break easily when being woven into fabrics. To prevent this breakage, they are coated with a jelly-like substance through a process called sizing. However, after the threads have been woven into fabrics, the agents needed to further finish the material cannot adhere to the jelly-coated fabrics. Thus, the protective sizing agents must be removed by a process called de-sizing. Amylase enzymes are widely used in de-sizing, as they do not weaken or affect cotton fibres, nor do they harm the environment.

- Retting of flax – Flax plants are an important source of textile fibres. Useful flax fibres are separated from the plant's tough stems through a process called retting. Traditional retting methods consume large quantities of water and energy. Bacteria, which may be bred or genetically engineered to contain necessary enzymes, can be used to make this a more energy efficient process.
- Breakdown of hydrogen peroxide – When cotton is bleached, a chemical called hydrogen peroxide, which can react with other dyes, remains on the fabric. Catalase enzymes specifically break down hydrogen peroxide and may be used to remove this reactive chemical before further dyeing.
- Biostoning and Biopolishing – Instead of using abrasive tools like pumice stones to create a stonewashed effect or to remove surface fuzz, cellulase enzymes may be used to effectively stonewash and polish fabrics without abrasively damaging the fibres.
- Detergents – Enzymes allows detergents to effectively clean clothes and remove stains. They can remove certain stains, such as those made by grass and sweat, more effectively than enzyme-free detergents. Without enzymes, a lot of energy would be required to create the high temperatures and vigorous shaking needed to clean clothes effectively. Enzymes used in laundry detergents must be inexpensive, stable, and

safe to use. Currently, only protease and amylase enzymes are incorporated into detergents. Lipase enzymes break down too easily in washing machines to be very useful in detergents. However, their stability is being studied and further developed through methods such as genetic screening and modification.

Synthetic fibres made from renewable sources of biomass are environmentally sustainable, and are becoming increasingly economically sustainable. Biodegradable synthetic polymers include novel fibres such as polyglycolic acid and polylactic acid, which are made from natural starting materials. Not all novel fibres are synthetic; they may also be naturally derived. Some natural biological fibres come from basic materials found in nature, including:

- Chitin – a type of sugar polymer found in crustaceans
- Collagen – a type of protein found in animal connective tissue
- Alginate – a type of sugar polymer found in certain bacteria

A prime example of a synthetic biomass fibre is Polylactic Acid (PLA), which is made by fermenting cornstarch or glucose into lactic acid, and then chemically transforming it into a polymer fibre. With properties similar to other synthetic fibres, PLA based materials are durable with a silky feel, and may be blended with wool or cotton.

PLA has potential applications in several areas, including the following:

- Textiles – clothing, fashions, upholstery
- Agriculture – plant mats, tree nets, soil erosion control products

- Sanitation – household wipes, diaper products
- Medicine – disposable garments, medical textiles

PLA minimizes environmental waste, as it may be fully biodegraded by microorganisms under appropriate conditions into carbon dioxide and water. Unlike the non-renewable petroleum resources used to make traditional synthetic fibres, the supply of renewable corn biomass needed to make PLA is expected to surpass demand in the anticipated future.

Biodegradable synthetic fibres and natural biological fibres may be used to make textiles for medical applications. Such textiles may be used in first aid, clinical, and hygienic practices. Some examples are described below:

Polymer Use(s)

Polylactic Acid and Polyglycolic Acid Used in sutures, absorbable wound closure products, orthopaedic repair absorbable pins, and fixation devices, as well as in tissue engineering structures

Collagen Uses in cell engineering structures, such as in artificial skin, or even as surgeon´s thread

Synthetic biomass fibres may also be used in drug delivery systems, which are designed to release drugs at a specified rate for a specified time.

(d) Enzymes in Industrial Applications

Enzymes are proteins that speed up biochemical reactions without being consumed or changed by the reaction. They are found throughout nature; in our bodies, in the environment, and in all living things. Without enzymes, life would not be possible. Enzymes speed up chemical and biochemical reactions, and this process is called catalysis. The substances upon which enzymes work when performing catalysis are known as substrates, and each enzyme will only fit and act upon a specific set of substrates.There are so many enzymes that it would be impossible to name them all. In fact, scientists have yet to discover many enzymes, or fully understand their structure and properties. On the other hand, many other enzymes have been successfully studied and applied to industrial and commercial uses. A few basic enzyme types are briefly described as follows:

Type	*Enzyme Function*
Cellulase	Breaks down cellulose, a fibre found in the cell walls of all plants and trees. Cellulose is the basic raw material used to make products such as paper, cotton, and other textiles
Hemicellulase	Breaks down hemicellulose, another plant sugar that is not as complex as cellulose and is easier to break down
Xylanase	Breaks down xylan, a gummy sugar present in the cell walls of plants and trees. This enzyme type is used primarily in the wood and pulp industry
Amylase	Breaks down starches and other carbohydrates into basic sugars
Protease	Breaks down proteins
Lipases	Breaks down fats

In the absence of enzymes, chemical reactions occur only when molecules collide while in proper alignment with each other. Because molecules are bumping into each other randomly, chemical reactions are essentially due to chance events. This sometimes results in reactions that occur very slowly, or reactions that do not occur at all.Enzymes act like tiny molecular machines to ensure that molecules come into contact with each other and react. Like a key fitting into a lock, chemical molecules fit into pocket-like structures located on an enzyme. These pockets hold the molecules in a position that will allow them to react with each other, ensuring that they are close enough together and aligned properly for a reaction to occur.

In this way, enzymes speed up reactions. The enzymes are not changed themselves by the reaction. When the reaction is complete, enzymes release the product(s) and are ready to bring together more molecules and catalyse more reactions. Biotechnology and EnzymesBiotechnology can provide an unlimited and pure source of enzymes as an alternative to the harsh chemicals traditionally used in industry for accelerating chemical reactions. Enzymes are found in naturally occurring microorganisms, such as bacteria, fungi, and yeast, all of which may or may not be genetically modified.

Large quantities of enzymes are often needed for industrial use, so these microorganisms are multiplied through a process called fermentation. When enzymes or the microorganisms used to create them are no longer needed, they are destroyed through exposure to heat or safe organic/ inorganic materials.

Enzyme Biotechnology in the Pulp and Paper Industry

Enzymes have been used in the pulp and paper industry to soften wood fibres, improve drainage, and present alternatives to chemical bleaching.

- Biopulping - Paper is made from cellulose fibres, which must be separated from a tough wood fibre called lignin. The step by step process used to separate cellulose from lignin and other wood components is known as pulping. It is a time and energy consuming process, involving the mechanical processing of wood or the treatment of wood with harsh chemicals. In biopulping, cellulase and xylanase enzymes made by lignin-degrading fungi are used to pre-treat wood and break down the lignin fibres. Removing lignin prior to further wood pulping saves time and energy, and decreases the quantities of chemicals used.

- Draining - Enzymes can also improve water drainage during wood pulping, a process that often slows down paper production. When fine lignin fibres are degraded by enzymes, less water is absorbed, thereby reducing drainage times, lessening the energy required to dry the paper, and producing a cleaner water runoff.
- Bleach Boosting - Lignin fibres that remain in wood pulp are coloured and must be bleached, usually by harsh chlorine compounds under high pressures. As an alternative, enzymes may be used to remove fine surface fibres, thereby reducing the bleaching process or eliminating it altogether.

Enzyme Biotechnology and Textiles

Enzymes are used to treat and modify fibres, particularly during textile processing and in caring for textiles afterwards. For example, enzymes called catalases are used to treat cotton fibres and prepare them for the dyeing processes. Some bacterial enzymes are used to separate the tough stem of the flax plant from the flax fibres used in textiles. By degrading surface fibres, many enzymes, including some cellulases and xylanases, are used to finish fabrics, give jeans a stonewashed effect, or help in the tanning of leathers. A recombinant enzyme called laccase, made by certain fungi, may also be used to treat fabrics and even catalyse the synthesis of some synthetic fibres. There are even enzymes in regular laundry detergent to help break down dirt, clean clothes more effectively, and prevent the dulling of fabric colours.

Enzymes are frequently used in laundry detergents. Without them, very high temperatures and mechanical shaking would be required to effectively clean clothes and other textiles.

Enzyme Biotechnology and Biofuels

Enzymes may be used to help produce fuels from renewable sources of biomass. Such enzymes include cellulases, which convert cellulose fibres from feedstocks like corn into sugars. These sugars are subsequently fermented into ethanol by microorganisms. A new process called Simultaneous Saccharification and Fermentation has greatly improved ethanol production efficiency. In this new process, cellulase enzymes and fermentation microorganisms are combined in a single reaction mixture to produce ethanol in one step, rather than producing sugars from cellulose and then fermenting them into ethanol separately.

Advances in science and technology have allowed researchers to improve enzymes through modifications to enzyme producing microorganisms, or through direct changes to the enzymes themselves.

- Genetic Engineering and Recombinant DNA technology - By using recombinant DNA technology, microorganisms may be genetically modified to produce a desired enzyme under specific conditions. This is accomplished through recombinant DNA technology, whereby small circular pieces of DNA, known as plasmids, are used to insert enzyme-producing genes into the genomes of organisms that possess another desirable trait, such as the ability to thrive on inexpensive nutrients. Therefore, both the enzyme and the original trait will be expressed in a single recombinant microorganism.

- Protein Engineering - All enzymes are made of proteins, which are large molecules formed from basic units, called amino acids, strung together like beads on a chain. To form functional enzymes, long chains of amino acids must be folded properly. Some enzymes consist of only one chain, whereas others are made of several chains that fit together. Scientists are using technology to study how proteins are formed, how they fold, and how they function. By studying the relation between the structure of a protein and how it functions, they are developing ways to improve and engineer enzymes. Proteins may be modified by changing one or more amino acids, and/or changing the way the amino acid chains fold and fit together.

Enzymes are a sustainable alternative to the use of harsh chemicals in industry. Because enzymes work under moderate conditions, such as warm temperatures and neutral pH, they reduce energy consumption by eliminating the need to maintain extreme environments, as required by many chemically catalysed reactions. Reducing energy consumption leads to decreased greenhouse gas emissions by power stations. Enzymes also reduce water consumption and chemical waste production during manufacturing processes. Because enzymes react specifically and minimize the production of by-products, they offer minimal risk to humans, wildlife, and the environment. Enzymes are both economically and environmentally feasible because they are safely inactivated and create little or no waste; rather than being discarded, end-product enzymatic material may be treated and used as fertilizer for farmers' crops. Bottom of Form

(e) Microbial Enhanced Oil Recovery

Microbial Enhanced Oil Recovery (MEOR) is the use of microorganisms to retrieve additional oil from existing wells, thereby enhancing the petroleum production of an oil reservoir. In this technique, selected natural microorganisms are introduced into oil wells to produce harmless by-products, such as slippery natural substances or gases, all of which help propel oil out of the well. Because these processes help to mobilize the oil and facilitate oil flow, they allow a greater amount to be recovered from the well.

MEOR is used in the third phase of oil recovery from a well, known as tertiary oil recovery. Recovering oil usually requires two to three stages, which are briefly described as follows:

- *Stage 1:* Primary Recovery – 12% to 15% of the oil in the well is recovered without the need to introduce other substances into the well.
- *Stage 2:* Secondary Recovery – The oil well is flooded with water or other substances to drive out an additional 15% to 20% more oil from the well.
- *Stage 3: Tertiary Recovery* – This stage may be accomplished through several different methods, including MEOR, to additionally recover up to 11% more oil from the well.

The microorganisms used in MEOR can be applied to a single oil well or to an entire oil reservoir. They need certain conditions to survive, so nutrients and oxygen are often introduced into the well at the same time. MEOR also requires that water be present. Microorganisms grow between the oil and the well's rock surface to enhance oil recovery by the following methods:

- Reduction of oil viscosity – Oil is a thick fluid that

is quite viscous, meaning that it does not flow easily. Microorganisms help break down the molecular structure of crude oil, making it more fluid and easier to recover from the well.

- Production of carbon dioxide gas – As a by-product of metabolism, microorganisms produce carbon dioxide gas. Over time, this gas accumulates and displaces the oil in the well, driving it up and out of the ground.
- Production of biomass – When microorganisms metabolize the nutrients they need for survival, they produce organic biomass as a by-product. This biomass accumulates between the oil and the rock surface of the well, physically displacing the oil and making it easier to recover from the well.
- Selective plugging – Some microorganisms secrete slimy substances called exopolysaccharides to protect themselves from drying out or falling prey to other organisms. This substance helps bacteria plug the pores found in the rocks of the well so that oil may move past rock surfaces more easily. Blocking rock pores to facilitate the movement of oil is known as selective plugging.
- Production of biosurfactants – Microorganisms produce slippery substances called surfactants as they breakdown oil. Because they are naturally produced by biological microorganisms, they are referred to as biosurfactants. Biosurfactants act like slippery detergents, helping the oil move more freely away from rocks and crevices so that it may travel more easily out of the well.

An Exopolysaccharide Called Xanthan: Case Study

The Xanthomans campestris bacteria produces a gummy

substance called Xanthan. Because Xanthan is molecularly composed of many different sugars and is externally secreted, it is known as an exopolysaccharide. Xanthan may be used in MEOR to lubricate oil drills, to help remove rocks from the drill site, and to compensate for decreased pressure in depleted oil wells, thereby facilitating the movement of oil up and out of the well.

Biotechnology and MEOR

MEOR is a direct application of biotechnology. It uses biological materials, such as bacteria, microorganisms, and their products of metabolism to facilitate the movement of oil out of a well, thereby enhancing oil recovery. Other applications of biotechnology in MEOR include genetic engineering techniques and recombinant DNA technology, which are used to develop strains of bacteria with improved oil recovery traits.

By inserting genes from one type of bacteria into another, scientists may combine two desirable genetic traits into one microorganism. For example, the temperature within an oil well is often too high for most microorganisms to survive. By inserting a gene that codes for a bacteria's ability to aid oil recovery into the genome of an existing bacteria that can survive under high temperatures, scientists may produce microorganisms that can both survive the heat of an oil well and also help retrieve oil. On their own, each bacteria lacks a trait necessary for oil recovery operations, but when combined through genetic engineering, the bacteria become integral to MEOR. The environmental conditions in an oil well make it very difficult for bacteria to survive, and those that do often have a decreased ability to carry out the chemical processes needed to enhance oil recovery. Researchers are working to create strains of bacteria that are better able to survive such harsh conditions but still retain the ability to carry out the chemistry needed for MEOR. Genetic engineering is being used to develop microorganisms that can not only live in the high temperatures of an oil well, but can also subsist on inexpensive nutrients, remain chemically active, and produce substantial amounts of biosurfactants. Some researchers are developing bacteria that can be grown on inexpensive agricultural waste material, which is abundant in supply and is environmentally friendly. As MEOR reduces or eliminates the need to use harsh chemicals during oil drilling, it is an environmentally compatible method of carrying out tertiary oil recovery. MEOR will become increasingly economically feasible as genetic engineering develops more effective microbial bacteria that may subsist on inexpensive and abundant nutrients. Methods for developing and growing MEOR bacteria are improving, thereby lowering production costs and making it a more attractive alternative to traditional chemical methods of tertiary oil recovery.

(f) Starch Bioprocessing

Bioprocessing is the use of biological means, such as the action of enzymes or other microorganisms, to treat various materials using a set of defined procedures. When this process is used to treat starch-based biomass, it is referred to as starch bioprocessing. Starch is a complex carbohydrate, found in cereal crops like wheat and corn. Because starch-containing biomass forms the basis of many lubricants, fuel additives, and other biochemicals commonly sold on the market, starch bioprocessing is often used in product manufacturing, in the conversion of waste materials to useful by-products, and in the breakdown of waste materials.

Raw biomass from various organic sources, including wheat and corn, is the starting material used to make lubricants, fuel additives, and other biochemical products through bioprocessing. Because these products are commonly sold on the market, they are known as commodity biochemicals. The following steps outline the production of commodity biochemicals from a corn-based source of biomass:

- Pretreatment – A process called corn wet milling is used to clean and treat corn raw materials before they are used as feedstocks in bioprocessing. Corn kernels are cleaned, soaked in a warm acidic liquid, and milled to produce a pulpy substance composed of corn starch, protein, and oil. Pretreatment requires little energy, and all products of this step are used. For instance, much of the pulpy substance is made into sweeteners, such as glucose and corn syrup, while the corn starch is reserved to make commodity biochemicals.
- Hydrolysis and Bioprocessing – Hydrolysis is a general term used to describe a chemical reaction that uses water to break down a larger compound, such as starch, into two or more smaller compounds, like sugars. Traditionally, harsh acids were used to initiate or speed up hydrolysis reactions, but in starch bioprocessing, such acids have been replaced by environmentally friendly microorganisms or enzymes. Bacterial enzymes are well suited for industrial use since they are inexpensive to produce and are stable at various temperatures. In this manner, enzymes bioprocess starches into sugars, which can then be made into commodity biochemicals.
- Further Bioprocessing – By studying how microorganisms produce particular biochemicals, scientists will be able to identify the genes that control these functions. It may be possible, then, to use recombinant DNA technology or cell fusion to introduce such genes into microorganisms and

genetically engineer them to process biochemicals more effectively and efficiently. By creating better microorganisms, starch bioprocessing techniques may be improved.

Corn starch is made of a long repeating chain of many small sugar molecules. During a hydrolysis reaction, the links between these molecules are broken down, resulting in single units of sugar. These single units are referred to as dextrose, or corn sugar.

Biotechnology in Starch Bioprocessing

The use of biotechnology is essential in starch bioprocessing techniques. Such biotechnologies use microorganisms, enzymes, or other biological materials to break down starches into sugars and further process them into chemicals. Biotechnology may also be used to genetically modify these microorganisms to increase the efficiency of sugar and chemical production from starches.

By studying how microorganisms produce particular biochemicals, scientists will be able to identify the genes that control these functions. It may be possible, then, to use recombinant DNA technology or cell fusion to introduce such genes into microorganisms and genetically engineer them to process biochemicals more effectively and efficiently. By creating better microorganisms, starch bioprocessing techniques may be improved.

Using microorganisms and enzymes in biochemical processing increases efficiency and decreases waste production. Biological techniques and other renewable processes, such as the use of enzymes to convert biomass resources into commodity biochemicals, must be increasingly used if biochemical production methods are to be environmentally sustainable.

A genetically modified food is a food product containing some quantity of any genetically modified organism (GMO) as an ingredient. Some nations have very strong disagreement over genetically modified organisms. For example, the European Union and Japan have enacted labelling and traceability requirements for GM food products, while the United States does not believe these requirements are necessary.

Background

Although "biotechnology" and "genetic modification" commonly are used interchangeably, GM is a special set of technologies that alter the genetic makeup of such living organisms as animals, plants, or bacteria. Biotechnology, a more general term, refers to using living organisms or their components, such as enzymes, to make products that include wine, cheese, beer, and yogurt. Combining genes from different organisms is known as recombinant DNA technology, and the resulting organism is said to be "genetically modified," "genetically engineered," or "transgenic." GM products (current or in the pipeline) include medicines and vaccines, foods and food ingredients, feeds, and fibers. Locating genes for important traits—such as those conferring insect resistance or desired nutrients—is one of the most limiting steps in the process. However, genome sequencing and discovery programs for hundreds of different organisms are generating detailed maps along with data-analyzing technologies to understand and use them.

History

The first commercially grown genetically modified food crop was a tomato created by Calgene called the FlavrSavr. Calgene submitted it to the U.S. Food and Drug Administration for testing in 1992; following the FDA's determination that the FlavrSavr was, in fact, a tomato, did not constitute a health hazard, and did not need to be labeled to indicate it was genetically modified, Calgene released it into the market in 1994, where it met with little public comment. Considered to have a poor flavor, it never sold well and was off the market by 1997.

Subsequent genetically modified food crops included virus-resistant squash, a potato variant that included an organic pesticide that kills caterpillars, named after the bacterium that produces it, Bt (NB: the US Environmental Protection Agency classified the Bt potato as a pesticide, but required no labeling), strains of canola, soybean, corn and cotton engineered by Monsanto to be immune to their popular herbicide Roundup, and Bt corn.

There was a brief interlude where Monsanto flirted with introducing a technology called terminator into food crops, which produced plants that grew sterile seeds. Monsanto claimed this was necessary to protect their intellectual property rights, since they were licensing the technology to farmers, and would also have provided a measure of protection against volunteer corn carrying unwanted traits, a major concern that arose during the Starlink debacle.

Controversies over risks

In August 1998 widespread concern, especially in Europe, was sparked by remarks by a leading GM researcher (with 270 published scientific papers to his name), Dr Arpad Pusztai, regarding some of his research into the safety of GM food. In his experiments, rats fed on genetically modified potatoes had suffered serious damage to their immune systems and shown stunted growth. He was vilified by leading British politicians, other scientists and by the GM companies, not least because his remarks, in a television interview, preceded the scientific publication of his results. Neither his eminence in the field nor his previous enthusiastic support for GM food were enough to save his career. Dr Pusztai was forced into retirement and his research suspended, whilst the British government blocked efforts to repeat his experiments which would have proved or disproved his claims.

In May 2005, a leaked report from Monsanto showed that some of its own experiments were raising doubts over the safety of GM food, and in particular seriously called into question the regulatory doctrine of substantial equivalence - that GM food with similar proteins and toxins is deemed no different than conventional food, without further investigation of the effects of any other differences. In Monsanto's research, rats fed on a diet rich in genetically modified corn developed abnormalities to internal organs and changes to their blood, raising fears that human health could be affected by eating GM food.

Whilst for many the safety of GM food for human consumption has yet to be demonstrated, the main public concerns have been over the environmental impacts of crops grown for food or for animal feed. In March 2005 these concerns were strengthened when the largest farm-scale trial comparing the biodiversity impact of GM crops with equivalent conventional crops found a significant negative impact on wildlife from GM.

Public reaction

Public outcry about the undue influence that the terminator gene (preventing plants from producing seeds) would give to Monsanto, particularly in less developed nations where seed saving is more common (in developed countries farmers usually tend to use the 1st generation seeds anyway), led to its withdrawal. Awareness grew throughout the nineties and eventually produced a strong backlash against GM foods (discussed below), which were panned as "untested", "unlabeled" and "unsafe"; following this backlash, the International Rice Research Institute, with funding from the Rockefeller Foundation developed a strain of rice enriched with vitamin A through genetic modification, dubbed golden rice. Subsequently the biotech industry touted this as a boon to poor people suffering from Vitamin A deficiency, which can cause blindness. This was condemned by GM food opponents as a ploy and a public relations move. (See golden rice for more.) Many prominent environmental organizations, like Friends of the Earth and Greenpeace, currently consider the issue of the presence of GMOs in conventional food products to be a major issue - indeed Greenpeace has made it a centerpiece of their activism. In 2002, opponents placed a measure on the Oregon ballot that would have made that state the first to require labelling of GMO food. Complicating the issue, the majority of GM crops grown today are fed to animals, thereby indirectly affecting human food production.

Application

Transgenic crops are grown commercially or in field trials in over 40 countries and on 6 continents. In 2000, about 109.2 million acres (442,000 km^2) were planted with transgenic crops, the principal ones being herbicide- and insecticide-resistant soybeans, corn, cotton, and canola. Other crops grown commercially or field-tested are a sweet potato resistant to a US strain of a virus that affects one out of the more than 89 different varieties of sweet potato grown in Africa, rice with increased iron and vitamins, and a variety of plants able to survive extreme weather.

Between 1996 and 2002, the total surface area of land cultivated with GMOs has increased by a factor of thirty. Land producing GMO crops grew from 17,000 km^2 (4.2

million acres) in 1996 to 520,000 km² (128 million acres) in 2001. The value for 2002 was 145 million acres (587,000 km²) and for 2003 was 167 million acres (676,000 km²). Soybean crop represented 63% of total surface in 2001, maize 19%, cotton 13% and canola 5%. In 2004, the value was about 200 million acres (809,000 km²) of which 2/3 were in the United States.

Four countries represent 99% of total GM surface in 2001: United States (68%), Argentina (22%), Canada (6%) and China (3%). It is estimated that 70% of products on U.S. grocery shelves include GM products. In particular, Bt corn is widely grown, as are soybeans genetically designed to tolerate Monsanto's Roundup herbicide.

The US Agriculture Department estimated that 38 percent of the 79 million acres (320,000 km²) of corn planted in 2003 will be genetically engineered varieties as well as 80% of the 73.2 million acres (296,000 km²) soybeans. The Grocery Manufacturers of America estimate that 75% of all processed foods in the U.S. contain a GM ingredient.

Future Applications

On the horizon are bananas that produce human vaccines against infectious diseases such as Hepatitis B; fish that mature more quickly; fruit and nut trees that yield years earlier, and plants that produce new plastics with unique properties. The next decade will see exponential progress in GM product development as researchers gain increasing and unprecedented access to genomic resources that are applicable to organisms beyond the scope of individual projects.

Technologies for genetically modifying (GM) foods offer dramatic promise for meeting some areas of greatest challenge for the 21st century. Like all new technologies, they also pose some risks, both known and unknown. Controversies surrounding GM foods and crops commonly focus on human and environmental safety, labeling and consumer choice, intellectual property rights, ethics, food security, poverty reduction, and environmental conservation (see below for a summary of "GM Foods: Benefits and Controversies").

Policy around the world

In 2000, countries that grew 99% of the global transgenic crops were the United States (68%), Argentina (23%), Canada (7%), and China (1%). Although growth is expected to plateau in industrialized countries, it is increasing in developing countries.

United States

In the United States, genetically modified food is widely available and accepted by consumers. The Food and Drug Administration assists companies in testing the safety of GM foods, but this process is voluntary. Labeling food as GM or non-GM is also voluntary. Some environmentalist groups believe the U.S. should regulate GM food more closely, and have called for mandatory labeling and testing requirements. Although agribusinesses are not required to test the safety of GM foods any more than non-GM foods, they have a legal duty to ensure that all their products are safe for human consumption.

Interestingly, some Amish people have adopted GM crops, because they are more productive, allow for less intensive farming (less pesticides, etc.), and do not conflict with the Amish lifestyle.

European Union

In Europe, a series of unrelated food crises during the 1990s (e.g. the BSE (or 'mad cow' disease) outbreaks and foot and mouth disease) have created consumer apprehension about food safety in general, and eroded the public trust in government oversight of the food industry. This has further fueled widespread public concern about GMOs, in terms of environmental protection (in particular biodiversity), health and safety of consumers and the right to make an informed choice. The apprehension might also be due to the perceived novelty of GM foods, as well as cultural factors relating to food. The mishandling of the BSE crisis has left some consumers unwilling to consider "science" to be a guarantee of quality.

Although some claim genetically modified foods may even be safer than conventional products, many European consumers are nevertheless demanding that their "right to know" the content and origin of the food they consume be respected. In a context of local food surplus where current GM food has little added nutritional value, many European consumers are wondering why any risk should be taken. However, as a result of the high quantity of GMO crops, the presence of GM in imported food products (shipments of grain for food, feed and processing for example), is now thought inevitable and largely unavoidable, and usually not mentioned.

EU regulation

For these reasons, the marketing of GM food is regulated in a manner that helps to provide the necessary levels of safety, transparency and reassurance. At the beginning of the 2000's, European officials insisted that new regulations were needed to "restore consumer confidence" in the technology. These new regulations required strict labelling and traceability of all food and animal feed containing more than 0.5 % GM ingredients. Directives, such as directive 2001/18/EC, were designed to require authorisation for the placing on the market of GMO, in accordance with the precautionary principle. (see also Tax, tariff and trade).

One of the features of the European system is a comprehensive pre-market risk assessment, a system trying to provide means for products to be followed at each stage of their production and distribution, by both transmission of accurate information and labelling. This traceability is a means to implement post-market measures such as monitoring and withdrawals (recalls).

This system is not only limited to GMO products but should encompass any food product ultimately.

The original EU rules for labelling of GM products were limited to products where transformed DNA and/or transformed protein are detectable, not to products that have been produced from GMOs but no longer appears to contain modified DNA and/or proteins. New rules for tracebility and labelling which came into force in 2004 also require labelling of highly refined products made from GM indgredirents like oil and corn syrup, even though that the presence of recombinant DNA or protein cannot be proven. The labelling rules do not apply to products of microbial genetic engineering, so the cheese made with the help of GM-chymosin doesn't have to be labelled. Officials stress that while traceability facilitates the implementation of safety measures, where appropriate, it cannot and should not be considered as a safety measure.

In 1999, a 4 year ban was pronounced on new genetically modified crops. At the end of 2002, European Union environment ministers agreed new controls on GMOs could eventually lead the 25-member bloc to reopen its markets to GM foods. European Union ministers agreed to new labelling controls for genetically modified goods which will have to carry a special harmless DNA sequence (a DNA code bar) identifying the origin of the crops, making it easier for regulators to spot contaminated crops, feed, or food, and enabling products to be withdrawn from the food chain should problems arise. A series of additional sequences of DNA with encrypted information about the company or what was done to the product could also be added to provide more data. (see Mandatory labelling).

See Trade war over genetically modified food for more details on disputes and more recent developments between the United States and the EU arising from EU position on genetically modified organisms.

Japan

Japan, like Europe, maintains labelling standards for GM food products. Japanese demand and assistance has led to a small effort to set up separate processing facility for non-GM soybeans in the U.S.

China and other developing countries

China is currently a producer of GM cotton, research published in Science shows that Chinese farmers growing GM cotton use significantly less pesticides, reducing costs and improving farmer health. The Chinese government has also released safety certificates following field and laboratory testing allowing the cultivation of GM tomato, pimiento and a species of morning glory. Development of new GM crops for food is an active field of research in Chinese institutions. In March 2002, China introduced biosafety rules that demanded strict labelling, extensive documentation and government approval for food shipments. Under these new rules, all soybean shipments from the United States were briefly interrupted until interim safety certificates could be acquired. In 2004 the Chinese Ministry of Agriculture announced its intention to assess the safety of GM rice lines developed by Chinese institutions for insect, disease and herbicide resistance, with government approval the crops may be planted as soon as spring 2006.

Poor nations' agriculture officials are receiving training courses on GMO at the American Agriculture Department, with instruction in the WTO rules on GM products and benefits of biotechnology. U.S. industry groups are also providing "technical assistance" to fund initiatives that promote "science-based and transparent biotechnology regulations" in countries such as China.

Key Biotechnology firms

The top 10 publicly-traded biotechnology companies, ranked by 2003 sales, are:

- Amgen
- Genentech
- Serono
- Biogen Idec
- Chiron Corporation
- Genzyme
- MedImmune
- Pfizer
- Millennium Pharmaceuticals
- Applied Biosystems

Social considerations

While some *individuals* may benefit from non-therapeutic genetic engineering, there may be adverse *social* implications. Few resources – particularly those related to medicine and health care – are available to everyone, and allowing the most privileged to engineer themselves or their children to have special capabilities could lead to what some call a genetic aristocracy. Numerous enhancements via genetic engineering have been proposed, including increased memory, intelligence, and less need for sleep, in addition to some peoples' desires to alter their physical appearance. The advantages created by genetic engineering, either real or perceived, could lead to new forms of inequality between those with genetic enhancements and those without while also exacerbating current inequalities between rich and poor, men and women, and whites and communities of color. Other considerations include:

- Would society treat genetic engineered people differently?
- Would they be left behind, would they be considered second class humans?
- What if this created a different species of human, would they still be able to interbreed, would they want to?
- What place would genetically engineered humans and regular humans have in society?
- Could unequal access to genetic engineering lock in or exaggerate current class divisions?

Metaphysical considerations

- The metaphysical (or "spiritual") implications of genetically engineered humans are vast in scope; e.g., were individual personality shown to be exclusively the result of genetic information acted upon by the environment, the concepts of the human soul and free will could be proven specious.
- However, could this be shown by the success of actual attempts at genetic engineering any more (or less) than it is shown by what is already known about the roles of genes and the environment in shaping the phenotypical characteristics (including behavior) of living things?

Economic and political effects

- Many opponents of current genetic engineering believe the increasing use of GM in major crops has caused a power shift in agriculture towards Biotechnology companies gaining excessive control over the production chain of crops and food, and over the farmers that use their products, as well.
- Many proponents of current genetic engineering techniques believe it will lower pesticide usage and has brought higher yields and profitability to many farmers, including those developing nations. A few GM licenses allow farmers in less economically developed countries to save seeds for next year's planting.
- In August 2002, Zambia cut off the flow of Genetically Modified Food (mostly maize) from UN's World Food Programme. Although there were claims that this left a famine-stricken population without food aid, the U.N. program succeeded in replacing the rejected grain with other sources, including some foods purchased locally with European cash donations. In rejecting the maize, Zambians cited the "Precautionary Principle" and also the desire to protect future possibilities of grain exports to Europe.
- In December 2005 the Zambian government changed its mind in the face of further famine and allowed the importation of GM maize
- In April 2004 Hugo Chávez announced a total ban on genetically modified seeds in Venezuela.
- In January 2005, the Hungarian government announced a ban on importing and planting of genetic modified maize seeds, although these were authorized by the EU.

Genetic engineering (GE) is the manipulation of genetic material (ie, DNA or genes) in a cell or an organism in order to produce desired characteristics and to eliminate unwanted ones. GE includes a range of different techniques with many different uses, and can be applied to plants, animals and humans. For example, the genetic modification of food is a form of GE that involves manipulating the cells of plants such as maize, to increase the yields, make it more nutritious and to make it drought- and disease-resistant. However, the most contentious type of GE is definitely related to its applications in humans. GE in humans has opened up a Pandora's box of possibilities as it can be used for both the miraculous and the sinister. The cloning of Dolly the sheep in 1996 was a very important event. Until then, the cloning of a human was only possible in theory. Recent films and TV programmes such as *Gattaca, Mutant X, Dark Angel* and books such as *Brave New World* all focus on possible consequences of this technology – but what is the real deal?

Cloning

If you were told that there was a clone of you sitting in the next room, what would you expect the clone to look and act like? Probably, exactly the same as you. But, despite all the movies and TV programmes that have explored the possibility of exact clones, it is highly unlikely that a clone of you would look exactly the same and would certainly not act exactly the same. A clone is not completely genetically identical, as there are small differences in the genetic make-up just as there are with identical twins. Despite the fact that identical twins come from the same egg, after a while one begins to notice the differences between them in order to tell them apart. It has been discovered – by doing a number of studies on identical twins brought up in the same environment – that genes mysteriously react differently to the same environment. While cloning a whole human is certainly the ultimate challenge for genetic engineers, cloning is not limited to this goal. Cloning can be, and is, done on a much smaller scale and could involve no more than just the cloning of a single cell. Therefore, cloning can be divided into two types:

- reproductive cloning (which is the cloning of a whole organism); and
- therapeutic cloning (which is the cloning of cells or even organs or other tissue for transplant purposes).

Due to the fact that genetic differences are likely to exist between a clone and its donor, this uncertainty has led to many countries banning the reproductive cloning of humans.

Reproductive cloning

In order to make a clone of someone, one needs a living cell and a human egg (ovum). The nucleus of the egg, which contains the DNA, is removed and replaced with the nucleus from the cell of the person/animal to be cloned. A short electrical pulse then stimulates the egg to start dividing and the embryo is then implanted into the womb where it develops into a duplicate of the person that donated the cell nucleus. Clones created in this way are not 100% genetically identical, as there is some DNA from the original egg cell that is found outside the nucleus (mitochondrial DNA).

Therapeutic cloning and stem cells

In therapeutic cloning an embryo is created in the same way as reproductive cloning, but it is not implanted into the womb of a woman. Instead, *stem cells* are extracted after the embryo starts dividing in the first 14 days after fertilisation, which kills the embryo. Stem cells are special cells with the ability to reproduce and become one of 300 types of cells, eg, skin, liver cell, hair or blood cells. These cells are then used to grow the specific type of tissue or organ that is needed and has the advantage of being genetically identical to the patient who donated it, eliminating the problem of organ or tissue rejection. Currently, if someone has an organ transplant, there is quite a high possibility that their body will reject the foreign organ and so they not necessary have to suppress the immune system to lessen the chances of this happening. Stem cells could potentially be used to repair damaged or defective tissues around the body, such as the cells in the pancreas that stop producing insulin in diabetics.

Adult stem cells versus embryonic stem cells

The problem with embryonic stem cells is that many people feel that by using a human embryo and then killing it, you are actually killing a potential person. It is for this reason that the United States and other countries are calling for a global ban on all human cloning, including the use of embryonic stem cells. As a result, there has been new research into the potential of what is known as "adult stem cells". More than 30 years ago it was discovered that, in addition to being found in embryos, stem cells are also found in adults. Adult stem cells can be removed from a person without causing any harm. Until very recently, it was thought that adult stem cells were only suitable for cloning a few types of cells and tissues. However, new studies have found adult stem cells with almost the same abilities as embryonic stem cells in various human tissues, including: the spinal cord, the brain, connective tissue and in the blood of the umbilical cord. These studies have shown that it is possible for us to manipulate cells to take on new functions/start doing different jobs simply by placing them in different environments and "reprogramming" them. For example, neural (brain) stem cells in mice have been transformed into the blood stem cells that produce the different types of blood cells. In other animals, bone marrow stem cells have become brain cells and liver cells. In humans, bone marrow stems cells of a patient have been transplanted into the areas of the heart damaged by a heart attack, triggering new growth of heart tissue. Although adult stem cells are already being used successfully to treat a number of diseases, there are a number of problems to be overcome, such as finding and extracting the adult stem cells. Embryonic stem cells have not yet been used for therapy and research is still in the initial stages. Eventually, it may be that, rather than just one type, both types of stem cells will be used for different purposes – whichever proves to be the best for the situation. In the meantime, parallel research continues on both cell types.

Uses of Cloning

Although currently the risks of cloning outweigh the possible benefits, there are many different potential uses of human cloning technology:

- *Replacing organs and other tissues* – such as new skin for burn victims, brains cells for those with brain damage, spinal rod cells for the paralysed and complete new organs (hearts, liver, kidney and lungs). Pigs are also being genetically modified to make their organs more compatible with humans by removing the gene that causes rejection. People could have their appearance changed (cosmetic surgery) using their own cloned tissue and accident victims and amputees could also benefit from this tissue regeneration.
- *Infertility* – human cloning provides couples and individuals who are unable to have children with another potential option.

- *Replacement of a lost child* – parents who have lost a child through an accident or an illness could clone an identical "replacement" child.
- *Creating "donor" people* – cloned people could be created to provide a source of transplant material.
- *Gene therapy* – cloning technology could be used to prevent, treat and cure genetic disorders by changing the expression of a person's genes. This technology may also provide the cure for cancer by revealing how cells are switched on and off. Gene therapy could be used to treat somatic (body) cells where the change is not passed on to children, or germ (egg and sperm) cells where the changes are passed on.
- *Saving endangered species* – by boosting their numbers through creating clones. However, since clones are almost genetically identical, the genetic diversity of the species would not be increased.
- *Reversal of the ageing process* – once more is understood about the role that our genes play in the ageing process. However, some of the above uses carry with them some serious ethical implications.

Problems with Cloning Techniques

Dolly the sheep was created in 1996 using the cloning methods outlined above. Although Dolly was born looking normal, she has suffered from several problems associated with the cloning technique, including premature arthritis, which is thought to be a side-effect of the cloning. Other problems with the current cloning techniques, include:

Allowing Cloning

Around the world, different countries have different rules relating to cloning – some don't allow reproductive cloning, but do allow therapeutic cloning, while other countries allow both types. In South Africa currently, the Human Tissue Act prohibits both therapeutic and reproductive cloning of humans. This situation will change when chapters 6 and 8 of the National Health Act are promulgated - and therapeutic cloning will then be allowed under strict conditions, requiring Ministerial permission, but reproductive cloning on humans will remain strictly prohibited. In China, there are no laws against cloning either.

However, with the huge population problem and the policy of only one child per family, there is no interest in reproductive cloning. On the other hand, there is huge interest in therapeutic cloning and, as a result, there is a lot of it on the go in China. Some Chinese cultures believe that humans only become people when they participate in society – so, according to these cultures, embryos and foetuses are not considered to be human beings, thereby eliminating any ethical problems surrounding the creation and destruction of embryos to get the required stem cells.

The Chinese government is funding extensive research and drawing back many Chinese scientists from overseas, to undertake work they would not be allowed to do elsewhere. With easy access and no limits on obtaining the embryonic material they require, it is expected that Chinese scientists will race ahead of the rest of the world in therapeutic cloning technologies. In the UK there are clear rules banning reproductive cloning, but scientists in both the UK and Israel are allowed to generate new embryonic cell lines for therapeutic research. The law in Germany bans the extraction of stem cells from human embryos for research within the country, but in 2002 a new law was passed allowing some human embryonic stem cells to be imported. This means that German scientists are allowed to undertake work on embryonic stem cells if they originate from outside Germany.

In the United States there is no public funding for research on embryonic stem cells. Therefore, although there is no law against therapeutic cloning in the US, there is almost no public research happening due to the lack of money and access to the embryonic material. Simultaneously, an announcement was recently made stating that US public funds (about US$1,4 million, about R11,2 million) will be provided for studies using adult stem cells instead of embryonic stem cells. These different rules for different countries mean that if a scientist is banned from undertaking cloning in his/her home country, he/she can simply move to a country where it is allowed. To prevent this from happening, the US and about 30 other countries want a global ban on all forms of cloning. However, many countries don't agree and thus no progress is being made on implementing this ban.

Genetic engineering, genetic modification (GM), and the now-deprecated gene splicing are terms for the process of manipulating genes, usually outside the organism's normal reproductive process. It often involves the isolation, manipulation and reintroduction of DNA into cells or model organisms, usually to express a protein. The aim is to introduce new characteristics such as making a crop resistant to an herbicide, introducing a novel trait, or producing a new protein or enzyme. Examples include the production of human insulin through the use of modified bacteria, the production of erythropoietin in Chinese Hamster Ovary cells, and the production of new types of experimental mice such as the OncoMouse (cancer mouse) for research, through genetic redesign. Since a protein is specified by a segment of DNA called a gene, future versions of that protein can be modified by changing the gene's underlying DNA. One way to do this is to isolate the piece of DNA containing the gene, precisely cut the gene out, and then reintroduce (splice) the gene into a different DNA segment. Daniel Nathans and Hamilton Smith received the 1978 Nobel Prize in physiology or medicine for their isolation of restriction endonucleases, which are able to cut DNA at specific sites. Together with ligase, which can join fragments of DNA together, restriction enzymes formed the initial basis of recombinant DNA technology.

Genetic engineering, the use of various methods to manipulate the DNA (genetic material) of cells to change hereditary traits or produce biological products. The techniques include the use of hybridomas (hybrids of rapidly multiplying cancer cells and of cells that make a desired antibody) to make monoclonal antibodies monoclonal antibody, an antibody that is mass produced in the laboratory from a single clone and that recognizes only one antigen. Monoclonal antibodies are typically made by fusing a normally short-lived, antibody-producing B cell (see immunity) to a fast-growing cell, such as a cancer cell (sometimes referred to as an "immortal" cell).; gene splicing or recombinant DNA, in which the DNA of a desired gene is inserted into the DNA of a bacterium, which then reproduces itself, yielding more of the desired gene; and polymerase chain reaction polymerase chain reaction (pOl`-mYrâs') (PCR), laboratory process in which a particular DNA segment from a mixture of DNA chains is rapidly replicated, producing a large, readily analyzed sample of a piece of DNA; the process is sometimes called DNA amplification, which makes perfect copies of DNA fragments and is used in DNA fingerprinting DNA fingerprinting or DNA profiling, any of several similar techniques for analyzing and comparing DNA from separate sources, used especially in law enforcement to identify suspects from hair, blood, semen, or other biological materials found at the scene of a violent crime. Monoclonal antibody, an antibody that is mass produced in the laboratory from a single clone and that recognizes only one antigen. Monoclonal antibodies are typically made by fusing a normally short-lived, antibody-producing B cell (see immunity immunity, ability of an organism to resist disease by identifying and destroying foreign substances or organisms. Although all animals have some immune capabilities, little is known about nonmammalian immunity. Mammals are protected by a variety of preventive mechanisms, some of them nonspecific (e.g., barriers, such as the skin), others highly specific (e.g., the response of antibodies). Nonspecific Defenses) to a fast-growing cell, such as a cancer cell (sometimes referred to as an "immortal" cell). The resulting hybrid cell, or hybridoma, multiplies rapidly, creating a clone that produces large quantities of the antibody.

Monoclonal antibodies engendered much excitement in the medical world and in the financial world in the 1980s, especially as potential cures for cancer. They have been used in laboratory research and in medical tests since the mid-1970s, but their effectiveness in disease treatment has been limited. By the mid-1990s, however, some of the technical problems had been overcome. Experimental cancer therapies have used drugs, radioactive materials, or immune killer cells attached to monoclonal antibodies that, when injected into patients, home in on antigens that grow only on the surface of cancer cells.

Genetically engineered products include bacteria designed to break down oil slicks and industrial waste products, drugs (human and bovine growth hormones, human insulin, interferon), and plants that are resistant to diseases, insects, and herbicides, that yield fruits or vegetables with desired qualities, or that produce toxins that act as pesticides. Genetic engineering techniques have also been used in the direct genetic alteration of livestock and laboratory animals (see pharming pharming (fär'm-ng), the use of genetically altered livestock, such as cows, goats, pigs, and chickens, to produce medically useful products. In pharming, researchers first create hybrid genes using animal DNA and the human or other gene that makes a desired substance, such as a hormone.). Genetically engineered products usually require the approval of at least one U.S. government agency, such as the Dept. of Agriculture, the Food and Drug Administration, or the Environmental Protection Agency.

Because genetic engineering involves techniques used to obtain patents on human genes and to create patentable living organisms, it has raised many legal and ethical issues. The safety of releasing into the environment genetically altered organisms that might disrupt ecosystems has also been questioned. The discovery in 2001 of genetically engineered DNA in native Mexican corn varieties made concerns of genetic pollution actual, and led some scientists to worry that the spread of transgenes through cross-pollination could lead to a reduction in genetic diversity in important crops. Imports of genetically modified corn, soybeans, and other crops have been curtailed or limited in some countries, and the vast majority of such crops are grown in just a handful of nations. The Cartagena Protocol on Biosafety, which has been signed by more than 100 nations and took effect in Sept., 2003, requires detailed information on whether and how imported seeds, plants, animals, other organisms, and the like are genetically modified and permits a nation to bar those imports. The United States, however, is not party to the treaty.

When Hungarian engineer Karl Ereky coined the word biotechnology in 1919 to describe the production of goods from raw materials with the aid of living organisms, he was describing techniques that had been employed by human beings for centuries. This includes techniques such as selective breeding, fermentation, hybridization, phyto-pharmacology, vaccination or the use of biomass for energy production.

However, the term biotechnology has evolved to include modern techniques with new applications, to what is now known as modern biotechnology. Modern biotechnology encompasses techniques such as genetic engineering (which is the direct intervention on the genetic makeup of an organisms usually by introducing foreign DNA into its gene pool by means that would not occur naturally), tissue culture and others. These modern techniques also take considerably less time to achieve the desirable changes in living organisms and carry greater precision than traditional techniques. In short, genetic engineering is a specific subset as a part of the more general subject of biotechnology.

Human genetic engineering refers to the the controlled modification of the human genome. DNA provides the the genetic blueprint for all living organisms and can influence individuals' actions and abilities. With the advent of DNA research and the ability to change gene expressions, it is now possible that scientists may be able to change human capacities, whether they be physical, cognitive, or emotional. Human genetic engineering is still in its infancy, however, with current research generally restricted to animals or gene therapy. Healthy humans do not need gene therapy to survive, though it may prove helpful to treat certain diseases. Special gene modification research has been carried out on groups such as the 'bubble children' - those who's immune system do not protect them from the bacteria and irritants all around them. The first clinical trial of human gene therapy began in 1990, but (as of 2006) gene therapy is still experimental. Other forms of human genetic engineering are still theoretical, or restricted to fiction stories. Recombinant DNA research is usually performed to study gene expression and various human diseases. Some drastic demonstrations of gene modification have been made with mice and other animals, however: testing on humans is generally considered off-limits. In some instances changes are usually brought about by removing genetic material from one organism, and transferring them into another species. This method is known as recombinant genetics. There are two main types of genetic engineering. Somatic modifications involve adding genes to cells other than egg or sperm cells. For example. if a person had a disease caused by a defective gene, a healthy gene could be added to the affected cells to treat the disorder. The distinguishing characteristic of somatic engineering is that it is non-inheritable, e.g. the new gene would not be passed to the recipient's offspring. Germline engineering would change genes in eggs, sperm, or very early embryos. This type of engineering is inheritable, meaning that the modified genes would appear not only in any children that resulted from the procedure, but in all succeeding generations. This application is by far the more consequential as it could open the door to the perpetual and irreversible alteration of the human species. There are two techniques researchers are currently experimenting:

- Viruses are good at injecting their DNA payload into human cells and reproducing it. By adding the desired DNA to the DNA of non-pathogenic virus (a pathogen is something that harms you. A non pathogenic virus is still a virus by definition, but will not cause you to get sick), a small amount of virus will reproduce the desired DNA and spread it all over the body.
- Manufacture large quantities of DNA, and somehow package it to induce the target cells to accept it, either as an addition to one of the original 46 chromosomes, or as an independent 47th human artificial chromosome.

Genetic engineering is most easily accomplished by making changes just after the egg and sperm have melded but before first division. In this way, the gene will be expressed throughout and will affect the recipients children, grandchildren, and all subsequent generations. This type of germline engineering is highly controversial and deemed inappropriate by most scientists. As of now, this is likely to take the form of gene therapy. This is less controversial as the engineered person can give consent, unlike germline engineering where subsequent generations are brought into the world with a genetic modification made to their body without their consent. Each human cell contains a different set of genes, so different genes could be transferred to different parts of the body. Such changes will not be hereditary unless the sex cells are engineered.

Bioengineering is the branch of engineering science in which biological science is used to study the relation between workers and their environments. It deals with the application of engineering principles to the fields of biology and medicine, as in the development of aids or replacements for defective or missing body organs and is also called biomedical engineering.

Protein engineering is the application of science, mathematics, and economics to the process of developing useful or valuable proteins. It is a young discipline, with much research currently taking place into the understanding of protein folding and protein recognition for protein design principles. There are two general strategies for protein engineering. The first is known as *rational design*, in which the scientist uses detailed knowledge of the structure and function of the protein to make desired changes. This has the advantage of being generally inexpensive and easy, since site-directed mutagenesis techniques are well-developed. However, there is a major drawback in that detailed structural knowledge of a protein is often unavailable, and even when it is available, it can be extremely difficult to predict the effects of various mutations. Computational protein design algorithms seek to identify amino acid sequences that have low energies for target structures. While the sequence-conformation space that needs to be searched is large, the most challenging requirement for computational protein design is a fast, yet accurate, energy function that can distinguish optimal sequences from similar suboptimal ones. Using computational methods, a protein with a novel fold has been designed, as well as sensors for un-natural molecules.

The second strategy is known as directed evolution. This is where random mutagenesis is applied to a protein, and a selection regime is used to pick out variants that have the desired qualities. Further rounds of mutation and selection are then applied. This method mimics natural evolution and generally produces superior results to rational design. An additional technique known as DNA shuffling mixes and matches pieces of successful variants in order to produce better results. This process mimics recombination that occurs naturally during sexual reproduction. The great advantage of directed evolution techniques is that they require no prior structural knowledge of a protein, nor it is necessary to be able to predict what effect a given mutation will have. Indeed, the results of directed evolution experiments are often surprising in that desired changes are often caused by mutations that no one would have expected. The drawback is that they require high-throughput, which is not feasible for all proteins. Large amounts of recombinant DNA must be mutated and the products screened for desired qualities. The sheer number of variants often requires expensive robotic equipment to automate the process. Furthermore, not all desired activities can be easily screened for. Rational design and directed evolution techniques are not mutally exclusive; good researchers will often apply both. In the future, more detailed knowledge of protein structure and function, as well as advancements in high-throughput technology, will greatly expand the capabilities of protein engineering.

Bioengineered Foods

Very basically, food-related biotechnology is the process by which a specific gene or group of genes with desirable traits are removed from the DNA of one plant or animal cell and spliced into that of another. Such beneficial genes might come from animals, (friendly) bacteria, fish, insects, plants and even humans. In some instances, genes that create problems (such as the natural softening of a tomato) are simply removed and not replaced. Tomatoes, for example, are generally picked green and gas-ripened later because, during shipping, they would become soft, bruised and unmarketable. A bioengineered tomato, however, can be picked ripe and shipped without softening. The objective of food biotechnology is to develop insect- and disease-resistant, shipping- and shelf-stable foods with improved appearance, texture and flavor. Additionally, biotechnology advocates say that the process will produce plants that are resistant to adverse weather conditions such as drought and frost, thereby increasing food production in previously prohibitive climate and soil conditions. They also envision increasing nutrient levels and decreasing pesticide usage through biotechnology. On the other hand, critics argue that, because biotechnology is producing new foods not previously consumed by humans, the changes and potential risks relating to such things as toxins, allergens and reduced nutrients are unpredictable. They also worry that, because genetically altered foods are not required to be labeled, people with religious or lifestyle dietary restrictions might unintentionally consume prohibited foods. In answer to such concerns, the FDA has issued the following evaluation guidelines by which a bioengineered food will be judged for approval:

- Has the concentration of a plant's naturally occurring toxicant increased?
- Has an allergic element not commonly found in the plant been introduced?
- Have the levels of important nutrients changed?
- Have accepted, established scientific practices been followed?
- What are the effects on the environment?

Biotechnology and Education

Biotechnology is the use of biological systems — living things — to make or change products. It has been used for centuries in traditional activities like baking bread and making cheese. Traditional biotechnology involved for example, developing a new wheat variety with early ripening characteristics by cross-breeding different types of wheat until the desired characteristics and only those characteristics were present in the new wheat variety. The terms 'gene technology', 'genetic engineering' and 'genetic modification' mean the same thing, and refer to one type of modern biotechnology. Since the 1940s scientists have known that DNA — deoxyribonucleic acid — in the cells of all living things is like a blueprint that is passed from one generation to another. Genes are made of DNA. They contain coded instructions for proteins, which give living things their particular characteristics like hair and eye colour. All living organisms use DNA as the genetic code. Gene technology includes a range of techniques to copy genes and modify them so that they will work in a new host either to produce or reduce a product. It also involves transfer of the reconstructed gene to a new host. Using gene technology, scientists aim to introduce, enhance or delete particular characteristics of a living thing, depending on whether they are considered desirable or undesirable. As you go further into this website, you can read more about gene technology — its history, how it's done, its possible uses, the regulatory processes in place for gene technology, the debate surrounding gene technology and how this technology might affect you.

Gene Technology

Gene technology is a term that refers to a whole range of techniques for genetic analysis that depend on the direct manipulation of DNA, the material substance of heredity. The use of gene technology in biomedicine, agriculture, food production and processing is an issue that has evoked strong public interest and concern. The debate has focused on the use of genetically modified organisms (GMOs) in food, their safety and potential impact on the environment. The public better accepts medical and industrial uses of gene technology. The majority of Australians are eager for more quality information about the technology and its applications.

Most people think of gene technology as adding genes from other life forms to plants and animals. While this is certainly part of it, the technology is broader than that, and has many more uses. Gene technology is developing at a rapid rate and it will continue to revolutionise basic biological research and development. It provides the potential to improve our health, create a safer and more secure food supply, generate greater prosperity and attain a more sustainable environment. Gene technology is already providing new ways of preventing, treating and curing human and animal diseases; it is helping farmers improve agricultural production with less impact on the environment; and in the near future, it will allow better food products to be available to consumers at reduced cost. The technology has given us vital new products like human insulin for diabetes, interferon and other drugs for treating certain cancers, and vaccines against diseases like hepatitis B. Millions of human lives are protected by gene technology every day. Genes are made from a chemical called DNA (deoxyribonucleic acid) that is common to all forms of life on Earth. Genes carry information for making all the proteins required by all organisms. These proteins determine, among other things, how the organism looks, how well its body uses food or fights infection, and sometimes even how it behaves. DNA is made up of four similar chemicals (called bases and abbreviated A,T, C, and G) that are repeated millions or billions of times throughout a genome. A genome is the entire DNA in an organism, including its genes. The human genome, for example, has 3 billion pairs of bases. The particular order of As, Ts, Cs, and Gs is extremely important. The order is called DNA sequence and variations in sequence underlie all of life's genetic diversity. DNA sequence is the database determining how the cell is to make proteins and what functions the proteins can perform. Because all organisms are related through similarities in DNA sequences, insights gained from non-human genomes often lead to new knowledge about human biology, and insights from the genes of a bacterium, worm or bee might contribute to new solutions for the wheat farmer. Gene technology consists of tools and techniques that scientists can use to study, identify or modify the genes of living organisms.

Techniques and tools

Genetic modification

Over 20 years ago it was discovered that genes, and

parts of genes, could be extracted from DNA using protein "scissors" then copied, or cloned. The gene could then be equipped with genetic "switches" to turn it up or down and inserted back into a living organism. This is the basis of genetic modification. It allows the sequences of genes to be determined. Transferring single genes between different plants and animals, turning existing genes up or down, or removing a gene from its original position and placing it in a new position in the same organism are all referred to as *genetic modification (GM)*. The terms genetic engineering (GE) or genetic manipulation also mean the same thing. Plants, animals or microbes which have a new gene inserted into them are called genetically modified organisms (GMOs) or transgenics. The modified gene belongs to the new host but its sequence is altered for example, on the basis of specific information gleaned from studies of other organisms.

DNA sequencing

DNA sequencing is the process of determining the exact order of the chemical building blocks (bases) that make up DNA. It is a laboratory procedure that involves first breaking down the DNA into short pieces, followed by separating the individual fragments using a technique called gel electrophoresis. A bar code pattern of the DNA pieces is produced which can then be read by computer. An enormous volume of information on the genetics of organisms is being generated using this technique and is providing computer experts with a great challenge in handling the data.

Human Genome Project

The Human Genome Project (HGP), which began in 1990, was a 13-year international project involving Australia. Its goal was to discover the approximate 50,000 genes that make up the human genome, make them accessible for further biological study and to determine the complete sequence of the 3 billion DNA subunits. Knowledge gained through this research will allow scientists and doctors of the future to prevent or treat most diseases at the level of genes. While it is now common to screen infants for inherited genetic diseases, further understanding of the human genome will allow doctors to determine a person's chances of developing particular diseases later in life. They will then be able to give advice or treatment to prevent, slow or cure those diseases.

Cloning technology

A "clone" is a copy of a plant, animal, microorganism or gene derived from a single common ancestor gene, cell or organism. Identical twins are natural clones from the one fertilised egg. A plant cutting is a clone of the original plant. Some confusion arises because the word *clone* is also applied to genes. A gene is said to be cloned when its sequence is multiplied many times in a common laboratory procedure. The cloning of a gene is a step in determining its sequence and initiating many types of experiments to understand its function and biology. The possibility of human cloning, raised when Scottish scientists at the Roslin Institute created the much-celebrated sheep "Dolly" in 1996, aroused worldwide interest and concern because of its scientific and ethical implications. Dolly was produced by injecting a nucleus from a mammary cell into an egg cell without a nucleus and raising an animal from that egg by treating it as a fertilised egg. Dolly was therefore a genetic clone of the animal from which the nucleus was taken. The important difference from identical twins is that the donor was a mature sheep and could even be made the mother of the clone. Cloning of animals and plants can be used to develop more efficient ways to produce superior breeds of animals. Cloning also enables genes to be added (such as those for human proteins) to produce animals that can generate hormones and other pharmaceuticals in milk, eggs or other products. The governments of many nations involved in gene technology have banned the cloning and genetic modification of human individuals. However, cloning technology may soon be available for human benefit to produce whole organs or special tissues from single cells for transplant to humans. It should be possible to use cells from the patient's own body and correct genetic defects before growing the tissue for transplant back into the donor. This is called therapeutic cloning.

Gene marker technology

DNA probes in diagnosis

Scientists can use gene technology to locate and analyse single genes in a chain of many thousands. They can make gene probes that recognise DNA sequences associated with genetic diseases. The diagnostic tests involve analysing minute tissue samples collected from adults, or taken from embryos. This information can assist parents and doctors to reduce the incidence of crippling genetic defects, and treat other disorders much better.

Forensic DNA fingerprinting

No two individuals have the same genetic makeup, a fact that is now being used by forensic scientists to identify individuals from cell samples left at a crime site. "Forensic DNA fingerprinting" involves taking a blood or tissue sample from a person and using one of a number of gene technologies to generate patterns of DNA fragments separated by size. Those patterns look something like a long bar code and are dependent on the exact sequences in the sample. A match between the genetic fingerprint from the crime sample and the suspect has been admissible evidence for a number of years.

DNA markers in breeding

For centuries plants and animals have been selectively

cross-bred for desirable traits such as size, enhanced quality and pest resistance. Scientists can now use DNA information to identify desirable traits in organisms. These DNA markers are helping breeders to select superior strains of plants or types of animals for commercial production. The technique is similar to the forensic DNA fingerprint but where a particular DNA bar is known to be associated with a specific superior trait. By using these markers superior animals or plants can be identified easily and early and these individuals selected for further breeding to improve the herd or crop. CSIRO research on gene markers includes:

- markers in cattle for carcass weight, fat content, colour, meat tenderness and leanness. This information assists breeders to select superior types of animals.
- in fish farming, markers are helping breeders of super prawns select for desirable traits like size and colour, disease resistance and environmental suitability.
- in crop breeding, markers are being used to combine genes for disease resistance and quality characteristics, such as higher yield and bread making properties in wheat.

Most animals, plants and fish are still bred in the same way they have been for over 2000 years — but gene technology means that breeders can now pick the best parents with greater precision.

Transgenics

Because DNA is the same in all living things, researchers have found it possible to use genetic information derived from one organism in the modification of the genes of another. It is possible to modify the physical characteristics in precise ways. The term 'transgenics' often used to describe transferring genes between species, but in reality the process is a sharing of genetic knowledge for specific outcomes. In practice, the modified gene often has some of the gene sequences and signals from the host with some sequence information provided by the donor. Gene transfers typically involve the exact modification of one gene while leaving the other 50,000 genes unaffected. Over the last 25 years, researchers have learned how to make precise genetic changes by identifying and transferring individual genes to produce a desired characteristic. Scientists can insert new genes into an organism to add a particular trait not found in that organism. The new organism is then tested over several generations before researchers know whether the modified organism has all the intended benefits. Methods for inserting a gene into an animal or plant include:

- microinjecting into an animal egg using a fine tube
- using harmless viruses to carry genes into a host
- inserting DNA into plant cells on gold or tungsten particles
- using *Agrobacterium,* a bacterium that naturally transfers DNA into plants.

Genes are being introduced into plants to produce:

- resistance to insect attack or microbial damage, so reducing pesticide sprays and natural carcinogenic aflatoxins (mould toxins)
- tolerance to benign herbicides, so weeds can be controlled and erosion minimised through minimum tillage farming
- higher quality protein and enhanced vitamin or iron levels
- healthier vegetable oils and starches.

Genes are being inserted into animals to make them:

- produce more meat and less fat
- grow faster
- produce more meat or milk for the same amount of feed
- become more resistant to disease and parasites.

Because the techniques used to transfer genes sometimes have a low success rate, before a gene is transferred, another gene — called a "selectable marker gene" - is sometimes attached to the target gene to let researchers know whether the new gene is present or not. Scientists often use genes for antibiotic-resistance as selectable markers to show if plants have taken up a new gene. This practice allows only the genetically modified cells to grow in a special culture that contains the specific antibiotic. The antibiotics are used only in the laboratory and are of no medical significance. The antibiotic resistance genes employed are already widely distributed through the natural environment, such that their presence in the plant is of no environmental consequence. Nevertheless, because of public concerns about antibiotic-resistance genes, researchers have developed ways to remove these gene markers from plants before they are released commercially. New types of marker genes have also been developed, such as those permitting growth in the laboratory on unusual sugars.

Gene silencing

A gene that is producing undesirable characteristics in an organism can be turned down or switched off. One way this can be achieved is by inserting a second copy of the gene, or a fragment of the gene, back to front. Other ways include RNAi and "gene shears", which cause the gene message molecule to destroy itself. This cancels the effect of the undesirable gene. Gene silencing is being used to:

- protect plants from viruses
- alter the colour of ornamental flowers
- delay ripening of tomatoes and peaches so they reach the consumer in better condition
- remove allergens (proteins that cause allergies) from peanuts, soybeans and wheat
- stop undesirable browning in potatoes and raw sugar.

Gene therapy

Many human and animal diseases have now been traced to their host being born with faulty genes that produce defective proteins. These diseases include sickle cell anaemia, cystic fibrosis, the blood clotting disorder haemophilia and Down's syndrome. Modern medicine and surgery have made great strides in treating patients with genetic diseases. Scientists are now exploring the use of gene therapy to attack these diseases at their source — by correcting faulty genes. Gene therapy holds great promise for treating disease by replacing or changing a very small part of the overall genetic program of carefully selected cells, perhaps permanently, producing a cure. It aims to restore the healthy function of cells by replacing or correcting the defective gene.

Gene therapy can be used to replace an abnormal gene with a normal one, to insert a missing gene, to switch off rogue genes that may cause cancers and to stop viruses multiplying within cells. The modified cells and genes are not passed onto children.

Certain technical obstacles are yet to be overcome if gene therapy is to deliver the expected benefits to treat human disease. Among the obstacles is the lack of a delivery system, or vector, that can safely and efficiently shuttle beneficial genes into the cells of patients — and ensure they work. Even the most advanced cell therapy techniques are still at the experimental stage in clinical trials. Further research is needed to develop safe, reliable gene therapy techniques.

- We could choose to have changes made to us, but we might also be making the choice for our children if the changes are carried through to the germline. Do we have that right, and how far should we take our ability?
- Conversely, is it responsible and ethically acceptable to leave the potentials of our children to the chance effects of the "genetic lottery", if we obtain the technological capacity to make positive changes?
- If genetic engineering became the way of the future, would people whose parents could not afford to genetically 'modify' them while still in an embryo, have a chance of achieving with high standards compared to the people who were 'modified' to be perfect?
- Is it ethical to experiment on embryos that have yet to be born?
- How would genetic engineering be used to revolutionize warfare?
- Who decides which changes will be made?

Today, women are faced with a rapidly expanding array of reproductive technologies. Developed by private biotechnology companies and marketed to fertility clinics, these new options have been presented to women as an issue of "choice"—supposedly providing them greater control over the process and outcome of their pregnancies. The Council for Responsible Genetics unequivocally supports a woman's right to make her own reproductive decisions.

However, we oppose the utilization of human eggs and embryos for experimental manipulations and as items of commerce because of the potential for eugenic applications and health risks to women and their offspring. The realm of assisted reproduction has become a multi-billion dollar industry, visible in the increasing availability of vitro fertilization, prenatal genetic diagnosis, and chemical and chromosomal testing of the fetus. So far, these experimental procedures have not been closely regulated. Part of this is because the reach of federal oversight extends exclusively to publicly- funded research, leaving private sector activity largely unregulated. Many are resistant to any regulations over the fertility industry due to the continuing political residue of the abortion debate. Women in the United States fought hard to have courts recognize a women's right to choose abortion, and are wary of opening the door to government regulation of anything to do with women's eggs, embryos or fetuses. The word "choice" has powerful connotations and women's groups themselves are not eager to restrict its power. The industry is aware of this and has capitalized on the situation by supposedly offering the new genetic technologies as "choices." The unique role of women in reproduction puts them on the front line of bio-technological experimentation, and as such, women have the potential to play a leading role in determining the direction and scope of these developments. CRG works, in partnership with a number of women's health groups, to reshape the discussion about genetic technologies in reproduction and to equip women with the information necessary to lead the process. Our goal is to shift debates over genetic technologies away from abortion politics and into more effective and productive discussions about the integrity of reproduction and the control of women's bodies. Through education, networking, outreach, and activism women can direct their own reproductive futures.

Agricultural Biotechnology

1.1 Biotechnology refers to any technique that uses living organisms, or parts of these organisms. Such techniques are used to make or modify products for a practical purpose. Modern medicine, agriculture, and industry make use of biotechnology on a large scale.

1.2 Traditional biotechnologies such as the use of yeast to make bread or wine have been applied for thousands of years. Since the late 19th century, knowledge of the principles of heredity gave farmers new tools for breeding crops and animals. They selected individual organisms with beneficial characteristics and developed hybrid crops.

New methods have been developed since the discovery of the DNA structure in 1954. For instance, micro-organisms can be used to produce antibiotics, and the hereditary material in plants can be changed to make them resistant to pests or diseases. .

2.1 Genes are the pieces of DNA code which regulate all biological processes in living organisms. The entire set of genetic information of an organism is present in every cell and is called the genome.

2.2 The genetic material is structured in a similar way in different species, which makes it easier to identify potentially useful genes. Certain species of crops, livestock, and disease-causing organisms have been studied as model species because they help us understand related organisms.

2.3 Certain fragments of DNA that can be easily identified are used to 'flag' the position of a particular gene. They can be used to select individual plants or animals carrying beneficial genes and characteristics. Important traits such as fruit yield, wood quality, disease resistance, milk and meat production, or body fat can be traced this way. More...

2.4 Plants can be obtained from small plant samples grown in test tubes. This is a more sophisticated form of the conventional planting of cuttings from existing plants. Another laboratory technique, in vitro selection, involves growing plant cells under adverse conditions to select resistant cells before growing the full plant. More...

2.5 In conventional breeding half of an individual's genes come from each parent, whereas in genetic engineering one or several specially selected genes are added to the genetic material. Moreover, conventional plant breeding can only combine closely related plants.

Genetic engineering permits the transfer of genes between organisms that are not normally able to cross breed. More...

2.6 For example a gene from a bacterium can be inserted into a plant cell to provide resistance to insects. Such a transfer produces organisms referred to as genetically modified (GM) or transgenic.

3. Does conventional plant breeding have effects on health and the environment?

In conventional plant breeding, little attention has been paid to the possible impacts of new plant varieties on food safety or the environment. Nonetheless, this kind of breeding has sometimes caused negative effects on human health. For instance, a cultivated crop variety created by conventional cross breeding can contain excessive levels of naturally occurring toxins.

The introduction of genetically modified plants has raised some concerns that gene transfer could occur in the field between cultivated and wild plants and such concerns also apply to conventional crops. Such transfers have occasionally been reported but are generally not considered a problem.

4. Are genetically modified plant foods safe to eat?

4.1 Foodstuffs made of genetically modified crops that are currently available (mainly maize, soybean, and oilseed rape) have been judged safe to eat, and the methods used to test them have been deemed appropriate. These conclusions represent the consensus of the scientific evidence surveyed by the International Council for Science (ICSU) and are consistent with the views of the World Health Organization (WHO).

However, the lack of evidence of negative effects does not mean that new genetically modified foods are without risk. The possibility of long-term effects from genetically modified plants cannot be excluded and must be examined on a case-by-case basis. New techniques are being developed to address concerns, such as the possibility of the unintended transfer of antibiotic-resistance genes.

Genetic engineering of plants could also offer some direct and indirect health benefits to consumers, for instance by improving nutritional quality or reducing pesticide use. More...

4.2 Scientists recommend that food safety assessment should take place on a case-by-case basis before genetically modified food is brought to the market. In such assessments, foodstuffs derived from genetically modified plants are compared to their conventional counterparts, which are generally considered safe due to their long history of use. This comparison considers to what extent different foodstuffs can cause harmful effects or allergies and how much nutrients they contain. More...

4.3 Consumers may wish to select foods on the basis of how they are produced, because of religious, environmental, or health concerns. However, merely indicating whether a product is genetically modified or not, without providing any additional information, says nothing about its content nor about possible risks or benefits. International guidelines are being developed for labelling genetically modified foods.

5. What effects could genetically modified crops have on the environment?

5.1 Agriculture of any type has an impact on the environment. Genetic engineering may accelerate the damaging effects of agriculture, have the same impact as conventional agriculture, or contribute to more sustainable practices.

Growing genetically modified or conventional plants in the field has raised concern for the potential transfer of genes from cultivated species to their wild relatives. However, many food plants are not native to the areas in which they are grown. Locally, they may have no wild relatives to which genes could flow.

Moreover, if gene flow occurs, it is unlikely that the hybrid plants would thrive in the wild, because they would have characteristics that are advantageous in agricultural environments only. In the future, genetically modified plants may be equipped with mechanisms designed to prevent gene flow to other plants.

A controversy has arisen about whether certain genetically modified plants (which are insect resistant because they carry the Bt gene) could harm not only insect pests but also other species such as the monarch butterfly. In the field, no significant adverse effects on non-target species have so far been observed. Nonetheless, continued monitoring for such effects is needed. More...

5.2 Genetically modified crops may have indirect environmental effects as a result of changing agricultural or environmental practices. However, it remains controversial whether the net effect of these changes will be positive or negative for the environment. For example, the use of genetically modified insect-resistant Bt crops is reducing the volume and frequency of insecticide use on maize, cotton and soybean. Yet the extensive use of herbicide and insect resistant crops could result in the emergence of resistant weeds and insects. More...

5.3 The broad consensus is that the environmental effects of genetically modified plants should be evaluated using science-based assessment procedures, considering each crop individually in comparison to its conventional counterparts.

6.1 Animal feeds frequently contain genetically modified crops and enzymes derived from genetically modified micro-organisms. There is general agreement that both modified DNA and proteins are rapidly broken down in the digestive system.

To date no negative effects on animals have been reported. It is extremely unlikely that genes may transfer from plants to disease-causing bacteria through the food chain. Nevertheless, scientists advise that genes which determine resistance to antibiotics that are critical for treating humans should not be used in genetically modified plants.

6.2 As of 2004, no genetically modified animals were used in commercial agriculture anywhere in the world, but several livestock and aquatic species were being studied. Genetically modified animals could have positive environmental impacts, for example through greater disease resistance and lower antibiotic usage. However, some genetic modifications could lead to more intensive livestock production and thus increased pollution.

7. Are GMOs regulated by international agreements?

7.1 Certain barriers to international agricultural trade have been reduced by the World Trade Organization (WTO). A WTO agreement adopted in 1994 establishes that countries retain their right to ensure that the food, animal, and plant products they import are safe. At the same time it states that countries should not use unnecessarily stringent measures as disguised barriers to trade. More...

7.2 Several international agreements relate to the environmental aspects of genetically modified crops. The Convention on Biological Diversity is mainly concerned with the conservation and sustainable use of ecosystems but also with environmental effects of GMOs. A part of this convention is the Cartagena Protocol on Biosafety, which regulates the export and import of genetically modified crops.

The International Plant Protection Convention was adopted to prevent the spread of pests affecting plants and plant products. It identified potential pest risks related to GMOs that may need to be considered, such as the potential development of invasive species or effects on beneficial insects and birds.

8. Conclusions

8.1 Agricultural biotechnology can be seen as both:

- a scientific complement to conventional agriculture, aiding for instance plant breeding programs, and
- a dramatic departure from conventional agriculture, enabling transfer of genetic material between organisms that would not normally mix.

Agricultural biotechnology has international implications and may become increasingly important for developing countries.1 However, research has tended to focus on crops important to developed countries. More...

8.2 To date, countries where genetically modified crops have been introduced in fields, have reported no significant health damage or environmental harm. Moreover, farmers are using less pesticides or using less toxic ones, reducing harm to water supplies and workers' health, and allowing the return of beneficial insects to the fields. Some of the concerns related to gene flow and pest resistance have been addressed by new techniques of genetic engineering.

However, the lack of observed negative effects does not mean that they cannot occur. Scientists call for a cautious case-by-case assessment of each product or process prior to its release in order to address legitimate safety concerns.

"Science cannot declare any technology completely risk free. Genetically engineered crops can reduce some environmental risks associated with conventional agriculture, but will also introduce new challenges that must be addressed. Society will have to decide when and where genetic engineering is safe enough." (FAO 2004)

9. Other views and links

This summary is based on the chapters 2 and 5 of "The State of Food and Agriculture 2003-2004; Agricultural Biotechnology" of the FAO (Food and Agriculture Organization), a leading scientific report produced by a large international panel of scientists.

It is considered by most scientists as a consensus document and other recent scientific assessments reach similar conclusions, though some people and organizations put forward different views.

2

Modern Advances in Biotechnology Science

In moder age biotechnology is any technique that uses living organisms or substances derived from these organisms to make or modify a product, improve plants or animals or develop microorganisms for specific uses. It has found diverse application in medicine, agriculture, food processing, chemical compounds and environmental management. The term "modern biotechnology" is used to distinguish activities from traditional biotechnology, which included fermentation technologies, such as fish sauce, soyasauce and beer making. Modern biotechnology has developed rapidly since the elucidation of the structure of DNA, with subsequent major developments including the advent of recombinant DNA technology, advanced cell and tissue culture techniques and modern immunology. Modern biotechnology has already provided a number of benefits in health care, agriculture and other industries. Biotechnology-derived drugs [biopharmaceuticals] are now routinely used in medicine and over 25 industrial and food crops have been genetically modified. For example, gene technology has enabled the production of insulin that is identical to that produced in healthy individuals, replacing insulin derived from pigs in many applications. Novel therapeutic and diagnostic products for the prevention and treatment of thrombotic [blood clotting] disorders. In agriculture, biotechnology has produced biotic and abiotic-resistant crops, and foods with improved nutritional qualities. The development of pest and disease resistant crops has increased yields and reduced the application of agricultural chemicals. For example, Bt cotton varieties grown in China, USA and other countries. Golden rice [higher vitamin A content] is an example of food with improved nutritional qualities.

Scope

The completion of entire genome sequences of many organisms including human provides a major impetus to understanding the complex nature of living cells. The international Human Genome Project stimulated developments both in high-throughput DNA sequencing, and in powerful computational tools for sequence analysis. Recently, the challenge turns from identifying the genome sequences to understanding their function. The post-genomics era refers to as functional genomics, the ability to monitor simultaneously potentially all events, whether it be the expression of genes at the RNA or protein level, protein-protein interactions, all alleles of all genes that affect a particular trait, or all protein-binding sites in a genome. DNA arrays can be used, most prominently to measure levels of gene expression for tens of thousands of genes simultaneously. Worldwide biotechnology markets amounted US$ 43 billion in 2000. Among the markets, the value in food and agriculture based on biotechnology was US$ 20 billion. The global market for biotechnology is expected to grow at 12 to 20% per year. Biotechnology is having a major impact on almost all the major sectors of industry. Its relative importance in industrial production and processing is estimated below. From the beginning of "new" biotechnology, the pharmaceutical and agri-chemical industries are being radically transformed by the application of recombinant and cell technologies.

Industry	*Relative importance*
Biopharmaceuticals	+++
Specialty chemicals	+
Colorants, food additives, cosmetics	++
Vaccines	+/++
Edible and commodity crops [agri-food industry]	++/+++
Biological clean up [bioremediation]	++
Biomining/ bioleaching	++

* Number of [+] indicates relative importance [more=higher]

Commercial biotechnology is a very research-intensive activity and driven by state-of-the art and private venture capital. Biotechnology is long dominated by the United States but it has now grown strongly, particularly in Germany, Britain, Israel and China. U.S. biotech firms total more than 1,300 compared with about 700 in all Europe. United States federal spending on biotechnology is US$ 6 billion annually, while many States have additional biotechnology programs. Canada Government expenditure on biotechnology research and development is estimated to be about C$ 300 million. An additional C$ 880 million was earmarked for a national genomics work. Germany has embraced biotechnology as key to its long-term competitiveness. In 1993 the government passed new legislation designed to streamline decision-making in biotech projects. In 1999, German researchers claimed 14% of all biotech patents applications, up from 10% five years ago and with more than 400 biotech-related start-ups now. Moreover, the country will channel $175 million over the next three years into a National Genome Research Network to work on the systematic functional analysis of genes and the use of those research results in the fight against widespread diseases. The new Genome Research Network involves at least 16 universities, several MaxPlanck institutes, and four national research centers. The Japanese Government spent US$ 2.5 billion in 1999 on biotechnology, an increase of 12.3% from the previous year. Major funding increases are focused on new research infrastructure in bioinformatics and genomics and on research commercialisation initiatives. Among the developing countries, the India government took the first step in encouraging a biotechnology industry back in 1986, by establishing a separate government department charged with increasing the number of biotechnology graduates coming from universities.

Fifty universities now produce about 500 biotech scientists annually. In addition, the government began funding more than 50 centers around the country to collect genomic data. China is reported to have 24 field tests and 20 commercialized of genetically engineered crops in 1999-2000. Comparing with the above mentioned countries, biotechnology in Thailand has grown at a lesser extent even though Thailand has substantial and unique genetic resources of plants, animals and microorganisms. Biotechnology can bring the enormous benefits and to maintain the competitiveness of Thailand's existing agricultural exports. A new era of biotechnology was started when The Thai Government established the National Center for Genetic Engineering and Biotechnology [BIOTEC] in 1983. Two years later, two specialized laboratories of BIOTEC, namely Plant Genetic Engineering and Microbial Genetic Engineering Units were initiated at Kasetsart and Mahidol University, respectively. Now, modern biotechnology is well developed in most universities and research institutes. The use of gene technology applications is wider and appreciated in the diagnosis of human and animal diseases.

But the barriers of biotechnology development are adequate and effective funding of public and private sector research and Thailand's ability to commercialise its research. Seventeen universities can now accept up to 600 biotechnology students annually. A survey by the Thai Society for Biotechnology showed that biotechnology scientists [with Bsc] have worked mainly in food and food beverage industries and only few with biotechnology industries. Co-ordination of research efforts across government portfolios is also important.

There are no existing mechanisms for such co-ordination which bring together all portfolios and public funding agencies with a biotechnology interest. Among the developing countries, the India government took the first stcp in encouraging a biotechnology industry back in 1986, by establishing a separate government department charged with increasing the number of biotechnology graduates coming from universities. Fifty universities now produce about 500 biotech scientists annually. In addition, the government began funding more than 50 centers around the country to collect genomic data. China is reported to have 24 field tests and 20 commercialized of genetically engineered crops in 1999-2000. Comparing with the above mentioned countries, biotechnology in Thailand has grown at a lesser extent even though Thailand has substantial and unique genetic resources of plants, animals and microorganisms. Biotechnology can bring the enormous benefits and to maintain the competitiveness of Thailand's existing agricultural exports. A new era of biotechnology was started when The Thai Government established the National Center for Genetic Engineering and Biotechnology [BIOTEC] in 1983. Two years later, two specialized laboratories of BIOTEC, namely Plant Genetic Engineering and Microbial Genetic Engineering Units were initiated at

Kasetsart and Mahidol University, respectively. Now, modern biotechnology is well developed in most universities and research institutes. The use of gene technology applications is wider and appreciated in the diagnosis of human and animal diseases. But the barriers of biotechnology development are adequate and effective funding of public and private sector research and Thailand's ability to commercialise its research. Seventeen universities can now accept up to 600 biotechnology students annually. A survey by the Thai Society for Biotechnology showed that biotechnology scientists [with Bsc] have worked mainly in food and food beverage industries and only few with biotechnology industries. Co-ordination of research efforts across government portfolios is also important. There are no existing mechanisms for such co-ordination which bring together all portfolios and public funding agencies with a biotechnology interest.

Biotechnology can be understood as the manipulation of living organisms to produce goods and services. Although the technology has received widespread media coverage in the past few years, it is a technology with a long history, dating as far back as 6000 B.C.

History

The term biotechnology was coined in 1919 by a Hungarian engineer called Karl Ereky. However, its origins date back further than that. The history of biotechnology can be divided into two eras: traditional and modern biotechnology. Traditional biotechnology dates back thousands of years, to early farming societies in which people collected seeds of plants with the most desirable traits for planting the following year. This practice is now known as selective breeding. The same selective breeding practices were used by early Babylonians, Egyptians, and

Romans to improve livestock. As far back as 6000 B.C., natural processes such as fermentation, in which microorganisms such as bacteria, yeasts and moulds play a critical role, were used to produce bread, beer and wine. Gregor Mendel's study of genetics, using seed and plant experiments at the end of the 19th century, gave the first indications of the cross from traditional to modern biotechnology. He discovered that traits are transmitted from parents to offspring by discrete, independent units, later called genes. His observations laid the groundwork for the field of genetics.

In 1943, the first direct evidence that deoxyribonucleic acid, or DNA, carried genetic information was discovered. However, it wasn't until 1953 that the mystery of the structure of DNA and the way genetic information is passed from generation to generation was unlocked by the discovery of the structure of DNA by James Watson and Francis Crick. The era of 'modern biotechnology', which involves manipulation of genes from living organisms in more precise and controlled ways than traditional biotechnology, began with their discovery. In 1985, genetically engineered plants resistant to insects, viruses, and bacteria were tested for the first time. Since then, many genetically engineered plants have been developed, successfully field tested and received food, livestock feed and environmental safety approval in many countries, including Canada. It was also in 1985 that a plan for mapping and sequencing the human genome was made. The goal of this Human Genome Initiative, which was launched in 1990, was to map all of the 80,000 to100,000 human genes by the year 2003.

Tools and Techniques

Biotechnology has many applications in health, agriculture, aquaculture and the environment. Health

Canada's role in biotechnology is to protect the health and safety of Canadians through regulation and public health policy. Biotechnology has been used by humans for thousands of years. For example, biotechnology has been used to make cheese, ferment wine and beer and make bread by using micro-organisms such as bacteria or fungi. Over time, we have also domesticated and selectively bred some animals and plants to meet human needs. In Canada, Health Canada, the Canadian Food Inspection Agency, Fisheries and Oceans Canada and Environment Canada, have joint responsibility to regulate biotechnology-derived products. The definition used is found in the Canadian Environmental Protection Act (1999): "the application of science and engineering to the direct or indirect use of living organisms of parts or products of living organisms, in their natural or modified forms."

Examples of biotechnology-derived products regulated by Health Canada are genetically modified and other novel foods, biologics and genetic therapies and assisted human reproduction technologies. Biotechnology can be best understood as an umbrella term referring to a variety of techniques, both traditional and modern. Note that some traditional techniques, such as selective breeding, hybridization and mutagenesis, are used in current applications of biotechnology, only with increased scientific knowledge and more advanced technology. These include:

- Selective breeding: The breeding of selected plants and animals to produce offspring with desired traits. The offspring with the desired traits are then used as breeding stock for the next generation and so on, until offspring that express the desired traits are obtained.
- Fermentation: The breakdown of complex organic

substances by micro-organisms, in the absence of oxygen. The process is energy-yielding.

- Hybridization: The production of offspring, known as hybrids, from genetically dissimilar parents. The object of hybridization is to combine desirable genes found in two or more different varieties to produce pure-breeding offspring superior in many respects to the parental types.
- Mutagenesis: The use of mutagens (such as exposure to radiation, temperature extremes and certain chemicals), to cause changes in the genetic make-up of cells, possibly resulting in new desirable, inheritable traits.
- Recombinant DNA (rDNA) techniques: The application of genetic techniques to produce desirable traits in living organisms using other living organisms such as bacteria. Examples of these techniques include the use of: restriction endonucleases (enzymes, produced by bacteria, that break foreign DNA molecules with the gene of interest into fragments which recombine with complementary molecules from a different source to form a recombinant DNA molecule); vectors (plasmids, often bacterial, or viruses that carry a piece of DNA into a bacterium for cloning purposes); and gene guns (called DNA particle guns and used to fire genes into cells by coating the genes onto tiny gold or tungsten particles and firing them from the barrel of the gun).
- Tissue Culture: A biological technique in which fragments of plant or animal tissue or cells are transferred to an artificial environment, free of other organisms, in which they continue to survive and

function for reproduction, chemical production and medical research.

Major changes in the field of biology have taken place over the last few decades. Advances in biotechnology and accomplishments such as the mapping of the entire human genome and the genomes of many other organisms are producing a better understanding of biological systems and living organisms. Large amounts of biological information and data have been and continue to be generated through this research. Bioinformatics arose from the need to organize, store, and analyse this information.

Bioinformatics is the use of computer software and programs to organize, store and analyse biological information and data to better understand biological systems. It is the field of science where biology, computer science and information technology converge as a single discipline. Researchers use advanced computer and statistical techniques to wade through and analyse large amounts of biological data created through modern biotechnologies referred to as the "omics" technologies – such as genomics and proteomics. These "omics" technologies focus on different aspects of biological systems. For example:

- Genomics is the analysis of the entire genetic make-up of an organism. It includes a study of the functions of genes in cells, organs and organisms.
- Proteomics focusses on the total protein make-up of a cell at any given time.

These "omics" technologies generate massive amounts of biological data through which it would be impossible to navigate without the use of computer systems. The data includes sequences of amino acids and nucleotides that underlie genes and proteins. The goal of bioinformatics is to translate the complex data gathered into usable knowledge. Bioinformatics typically includes three main areas:

- Developing new statistics and data to be studied and used to assess relationships among the different sets of biological data.
- Analysing and interpreting the various types of data, which include nucleotide and amino acid sequences and the properties of various proteins.
- Developing and implementing tools to enable efficient access and management of different types of pertinent information.

The field of bioinformatics emerged in the early 1980s with the creation of the GenBank database. GenBank, started by the US Department of Energy, stores DNA sequence information obtained from a wide range of organisms. In the early days, GenBank was a small scale operation with a roomful of technicians sitting at keyboards entering the DNA sequence information published in academic journals.

The advent of the Internet allowed researchers to access data in GenBank from all over the world for free. The DNA sequence data in GenBank grew rapidly with the emergence of highly sophisticated gene sequencing tools. Private companies joining the sequencing race with parallel projects created huge databases of their own.

Several services emerged as the need for access to bioinformatics increased. The two most significant were the European Molecular Biology Network (EMBnet) and the United States National Center for Biotechnology Information (NCBIO). In Canada, the Canada Institute for Scientific and Technical Information (CISTI) of the National Research Council of Canada (NRC) operated a bioinformatics server, Molecular Biology Database Service (MBDS), from

1987 to 1996. The Canadian Bioinformatics Resource (CBR) was created in 1995 to replace MBDS in providing bioinformatics resources and services to Canadian researchers.

Revolution in Agriculture Through Biotechnology

Biotechnology refers to any technique that uses living organisms, or parts of these organisms. Such techniques are used to make or modify products for a practical purpose. Modern medicine, agriculture, and industry make use of biotechnology on a large scale. Traditional biotechnologies such as the use of yeast to make bread or wine have been applied for thousands of years. Since the late 19th century, knowledge of the principles of heredity gave farmers new tools for breeding crops and animals. They selected individual organisms with beneficial characteristics and developed hybrid crops. New methods have been developed since the discovery of the DNA structure in 1954.

For instance, micro-organisms can be used to produce antibiotics, and the hereditary material in plants can be changed to make them resistant to pests or diseases. Genes are the pieces of DNA code which regulate all biological processes in living organisms. The entire set of genetic information of an organism is present in every cell and is called the genome. The genetic material is structured in a similar way in different species, which makes it easier to identify potentially useful genes. Certain species of crops, livestock, and disease-causing organisms have been studied as model species because they help us understand related organisms.

Certain fragments of DNA that can be easily identified are used to 'flag' the position of a particular gene. They can be used to select individual plants or animals carrying beneficial genes and characteristics. Important traits such as fruit yield, wood quality, disease resistance, milk and meat production, or body fat can be traced this way. Plants can be obtained from small plant samples grown in test tubes. This is a more sophisticated form of the conventional planting of cuttings from existing plants. Another laboratory technique, in vitro selection, involves growing plant cells under adverse conditions to select resistant cells before growing the full plant. In conventional breeding half of an individual's genes come from each parent, whereas in genetic engineering one or several specially selected genes are added to the genetic material. Moreover, conventional plant breeding can only combine closely related plants. Genetic engineering permits the transfer of genes between organisms that are not normally able to cross breed.

For example a gene from a bacterium can be inserted into a plant cell to provide resistance to insects. Such a transfer produces organisms referred to as genetically modified (GM) or transgenic.

In conventional plant breeding, little attention has been paid to the possible impacts of new plant varieties on food safety or the environment. Nonetheless, this kind of breeding has sometimes caused negative effects on human health. For instance, a cultivated crop variety created by conventional cross breeding can contain excessive levels of naturally occurring toxins. The introduction of genetically modified plants has raised some concerns that gene transfer could occur in the field between cultivated and wild plants and such concerns also apply to conventional crops. Such transfers have occasionally been reported but are generally not considered a problem.

Foodstuffs made of genetically modified crops that are currently available (mainly maize, soybean, and oilseed rape) have been judged safe to eat, and the methods used to test them have been deemed appropriate. These conclusions represent the consensus of the scientific evidence surveyed by the International Council for Science (ICSU) and are consistent with the views of the World Health Organization (WHO). However, the lack of evidence of negative effects does not mean that new genetically modified foods are without risk. The possibility of long-term effects from genetically modified plants cannot be excluded and must be examined on a case-by-case basis. New techniques are being developed to address concerns, such as the possibility of the unintended transfer of antibiotic-resistance genes. Genetic engineering of plants could also offer some direct and indirect health benefits to consumers, for instance by improving nutritional quality or reducing pesticide use. Scientists recommend that food safety assessment should take place on a case-by-case basis before genetically modified food is brought to the market. In such assessments, foodstuffs derived from genetically modified plants are compared to their conventional counterparts, which are generally considered safe due to their long history of use. This comparison considers to what extent different foodstuffs can cause harmful effects or allergies and how much nutrients they contain.

Consumers may wish to select foods on the basis of how they are produced, because of religious, environmental, or health concerns. However, merely indicating whether a product is genetically modified or not, without providing any additional information, says nothing about its content nor about possible risks or benefits. International guidelines are being developed for labelling genetically modified foods.

Agriculture of any type has an impact on the environment. Genetic engineering may accelerate the damaging effects of agriculture, have the same impact as conventional agriculture, or contribute to more sustainable practices. Growing genetically modified or conventional plants in the field has raised concern for the potential transfer of genes from cultivated species to their wild relatives. However, many food plants are not native to the areas in which they are grown. Locally, they may have no wild relatives to which genes could flow. Moreover, if gene flow occurs, it is unlikely that the hybrid plants would thrive in the wild, because they would have characteristics that are advantageous in agricultural environments only. In the future, genetically modified plants may be equipped with mechanisms designed to prevent gene flow to other plants. A controversy has arisen about whether certain genetically modified plants (which are insect resistant because they carry the Bt gene) could harm not only insect pests but also other species such as the monarch butterfly. In the field, no significant adverse effects on non-target species have so far been observed. Nonetheless, continued monitoring for such effects is needed.

Genetically modified crops may have indirect environmental effects as a result of changing agricultural or environmental practices. However, it remains controversial whether the net effect of these changes will be positive or negative for the environment. For example, the use of genetically modified insect-resistant Bt crops is reducing the volume and frequency of insecticide use on maize, cotton and soybean. Yet the extensive use of herbicide and insect resistant crops could result in the emergence of resistant weeds and insects. The broad consensus is that the environmental effects of genetically modified plants should be evaluated using science-based assessment procedures, considering each crop individually in comparison to its conventional counterparts.

Animal feeds frequently contain genetically modified crops and enzymes derived from genetically modified microorganisms. There is general agreement that both modified DNA and proteins are rapidly broken down in the digestive system. To date no negative effects on animals have been reported. It is extremely unlikely that genes may transfer from plants to disease-causing bacteria through the food chain. Nevertheless, scientists advise that genes which determine resistance to antibiotics that are critical for treating humans should not be used in genetically modified plants. As of 2004, no genetically modified animals were used in commercial agriculture anywhere in the world, but several livestock and aquatic species were being studied. Genetically modified animals could have positive environmental impacts, for example through greater disease resistance and lower antibiotic usage. However, some genetic modifications could lead to more intensive livestock production and thus increased pollution.

Certain barriers to international agricultural trade have been reduced by the World Trade Organization (WTO). A WTO agreement adopted in 1994 establishes that countries retain their right to ensure that the food, animal, and plant products they import are safe. At the same time it states that countries should not use unnecessarily stringent measures as disguised barriers to trade. Several international agreements relate to the environmental aspects of genetically modified crops. The Convention on Biological Diversity is mainly concerned with the conservation and sustainable use of ecosystems but also with environmental effects of GMOs. A part of this convention is the Cartagena Protocol on Biosafety, which regulates the export and import of genetically modified crops. The International Plant Protection Convention was adopted to prevent the spread of pests affecting plants and plant products. It identified potential pest risks related to GMOs that may need to be considered, such as the potential development of invasive species or effects on beneficial insects and birds. Agricultural biotechnology can be seen as both:

- a scientific complement to conventional agriculture, aiding for instance plant breeding programs, and
- a dramatic departure from conventional agriculture, enabling transfer of genetic material between organisms that would not normally mix.

Agricultural biotechnology has international implications and may become increasingly important for developing countries.1 However, research has tended to focus on crops important to developed countries. To date, countries where genetically modified crops have been introduced in fields, have reported no significant health damage or environmental harm. Moreover, farmers are using less pesticides or using less toxic ones, reducing harm to water supplies and workers' health, and allowing the return of beneficial insects to the fields. Some of the concerns related to gene flow and pest resistance have been addressed by new techniques of genetic engineering. However, the lack of observed negative effects does not mean that they cannot occur. Scientists call for a cautious case-by-case assessment of each product or process prior to its release in order to address legitimate safety concerns. "Science cannot declare any technology completely risk free. Genetically engineered crops can reduce some environmental risks associated with conventional agriculture, but will also introduce new challenges that must be addressed. Society will have to decide when and where genetic engineering is safe enough." (FAO 2004)

This summary is based on the chapters 2 and 5 of "The State of Food and Agriculture 2003-2004; Agricultural Biotechnology" of the FAO (Food and Agriculture Organization), a leading scientific report produced by a large international panel of scientists. It is considered by most scientists as a consensus document and other recent scientific assessments reach similar conclusions, though some people and organizations put forward different views.

Plant Transformation

The area of plant transformation should lead to the production of transgenic plants with superior properties including the resistance to diseases, insect pests and abiotic stress. The Plant Genetic Engineering Unit [PGEU], the specialized laboratory of BIOTEC at Kasetsart University, Kamphaengsaen Campus was established in 1985 to carry out the work on plant biotechnology and genetic engineering. Transgenic tomato plant carrying the coat protein gene of tomato yellow leaf curl virus was first developed to control the serious virus disease of tomato. The same approach was taken to develop transgenic papaya and pepper for the resistance to papaya ringspot virus and chilli vein-banding mottle virus, respectively. Development of transgenic rice varieties has been supported by the Rice Biotechnology Program launched by the BIOTEC and the Rockefeller Foundation. The example is transformation of Khaw Dawk Mali 105, an pyrroline-5-carboxylate synthetase [P5CS] for salt and?aromatic Thai rice with drought tolerance. Most of transgenic plants are now being tested under greenhouse conditions in accordance with the Biosafety Guidelines. The viral resistance papaya developed by Department of Agriculture is now undergoing field trial and being environmentally assessed for possible release.

Selection Process

Tomato production in the tropics and subtropics faces serious constraints due to bacterial wilt, a disease caused by the bacterial pathogen recently reclassified as Ralstonia solanacearum. In Thailand, an endemic outbreak of bacterial wilt [BW] in tomato, potato, pepper, ginger, and peanut occurs each year, causing a total yield loss of approximately 50-90% depending on growing conditions. BW resistant varieties cannot easily be accomplished due to the nature of the [quantitatively inherited] resistance that has several genes involved. Marker assisted selection [MAS]-a breeding method of selecting individuals based on markers linked to target genes in addition to phenotypic measurement only is essential and useful for enhanced resistance to diseases. Tomato germplasm containing resistance genes and the location of the genes of tomato molecular map allow the use of DNA markers located near the resistance genes as a tool in selection for the resistant traits. Currently, the breeding project for resistance tomato has conducted marker assisted selection not only for wilt disease caused by Ralstonia solanacearum but others such as, nematode, tobacco mosaic virus and powdery mildew.

There have been many approaches to agricultural development, however, few have been successful as well as sustainable. A long research effort started in the early 1950's resulted on the first officially approved maize variety, Suwan-1 which was released in Thailand in 1973. The research was initiated to overcome drought stress and other production constraints, most notably, sorghum downy mildew disease. Nowadays, downy mildew disease seems less severe in the farmer fields, but it is still the first rank of important diseases of maize in Thailand.

Breeding for downy mildew resistance is based on selection and hybridization of selected plants in screening nursery. The effectiveness of selection depends highly on environmental conditions of disease nursery. The molecular techniques such as Simple Sequence Repeat [SSR] are being used to define marker positions which link to QTLs of sorghum downy mildew for the application of breeding programs through marker assisted selection. Molecular genetic markers can now find direct application and should greatly assist for selective breeding programs of commercially important marine species in Thailand. Species-specific markers found in several marine species are used for identification of correct broodstock and seed species and for quality control of commercial trading of oyster, mud crab and abalone seed.

A population-specific RAPD marker found in black tiger prawn (Penaeus monodon) from the Andaman sea can be used to verify growth and survival performance among different black tiger prawn stocks in commercial culture settings. Due to powerful application of DNA-based technologies to identify DNA markers which are associated with nuclear loci that control economically important trait [quantitative trait loci, QTLS], marker assisted selection has great potential and gain more attention in breeding programs for plants and animals, for examples, rice resistance to bacterial leaf blight and leaf/neck blast.

Product Development

Thailand similar to other developing countries, needs to form network with other laboratories to facilitate technology transfer and get accessed to genetic material important in the construction of genetic modification. The construction of genetically modified organisms is usually intertwined with license fee and intellectual right obligations. For example, the construction of vitamin A enriched rice involves more than 70 licenses. Usually the reliance on the expertise of other collaborators means that strict restrictions are imposed on the use of the products. Capacity building and human resource development via research is a long-term solution for Thailand to become more independent in the product development using gene technology. BIOTEC as a research funding agency, is working actively towards this objective. One notable example is the US$ 3.5 million funding for the rice functional genomics project. BIOTEC on behalf of Thailand has joined an International Collaboration for Sequencing the Rice Genome [ICSRG]. This project will form network with researchers worldwide as well as within the countries. Joining ICSRG will allow Thai Scientists to directly access the rest of the genome sequence made available by the other collaborating members. Moreover, BIOTEC hopes to discover several important genes and regulatory elements such as strong inducible promoters for use to improve the genetic transformation process as well as the genetic improvement of rice. The project will bring Thailand to international scientific arena, incorporate state of the art technology and finally improve the competitive edge of Thailand in the international market.

Nucleic Acids, Proteins and Lipid Membranes

Perhaps the most specific biomolecular self-assembly occurs when two single strands of Deoxyribonucleic Acid (DNA) assemble into the infamous double helix. Based on Watson and Crick base pairing, two complimentary strands of DNA will find each other within an enormously complex mixture. Double helical DNA strands can be separated from one another through simple heating, resulting in single stranded DNA (ssDNA).

If the solution is allowed to cool, the two strands will re-assemble into a stable double helix. While the Watson and Crick base pairing then allows the specific assembly of the double helix, which is stabilized by relatively strong *hydrophobic* interactions between the aromatic bases within the center of the helix also known as *pi-stacking*. Nanoscientists can take advantage of the double helix's specific stability in numerous ways. One promising approach is to attach single-stranded DNA to small particles that one wishes to assemble. These components could range from diamondoid manifolds to metallic clusters that function as quantum dots. When DNA is used to create functional surfaces, the technique is called *DNA Directed Immobilization (DDI)*. Such patterned surfaces have applications for both nanoelectronic devices and biosensors.

Nanogen, a biosensor manufacturer has a versatile method for that takes advantage of DNA's negative charge. By applying a voltage to various parts of a DNA containing solution, the concentrations of DNA over various parts surfaces can be regulated. As the primary components of the molecular machinery that makes life possible, proteins have a vast array of resources available to them for biomolecular self-assembly. Take for instance the eukaryotic nucleus of a multicellular organism (such as a human). The organization of the information contained within this tiny compartment is so vast that scientists are only now beginning to attempt to understand it.

Containing the compartment is the nuclear membrane, which is filled with proteins that pump nutrients in and messages out. Entwined within what is called the "nuclear matrix" is the entire human genome, containing about 3 meters worth of DNA, packed into an area of only a few cubic micrometers. Keeping it all under control are then *histones* and thousands of other (non-histone) proteins that carry out processes such as *transcription* and *translation*. Each of these proteins has the appropriate information encoded within its peptide sequence for getting it to the location where it can do its job.

In order for each of our cells to produce those proteins required for its specific task, only certain genes can be transcribed in certain cells. Thus, a very important amount of information is contained not only in the sequence of the DNA, but in its spatial organization within the nucleus. For instance, when the nucleus divides, all the contents must be unraveled, copied, and then reassembled back into two daughters of the original organized state. All this is accomplished by means of biomolecular self-assembly.

Structural Proteins are generally called "fibrous" because they often have a fiber like nature. With names such as collagen, keratin, elastin, tubulin and fibroin, fibrous proteins self-assemble (often covalently) into large polymers that provide various nano- meso- and even macro-scale structures. Even our most modern synthetic polymers can not compete with fibrous proteins such as fibroin (silk) when it comes to strength, flexibility and overall molecular scale order. Thus, serious efforts are underway to better understand and take advantage of fibrous self-assembly. Enzymes (biological catalysts) are generally "globular" as opposed to "fibrous" and they often work together in the form of multi-enzyme complexes.

This means that both intramolecular self-assembly and intermolecular self-assembly must take place in order for the enzyme to reach the conformation that allows it to do its job. Intramolecular self-assembly means that the newly synthesized protein chain must fold into a functional state. Most often this is accomplished by the aggregation of hydrophobic residues in the center of the protein, thus forcing hydrophilic residues to the surface of the protein. Once a single protein chain is adequately folded, it will often begin intermolecular self-assembly. In order to assemble a molecular machine known as a "multienzyme complex," the numerous subunits (or proteins) contact one another in ways that cause new surfaces to appear, thus allowing further subunit components to bind. Since all the information for this self-assembly is often contained within the protein sequence, it is possible to reproduce this assembly in a test tube. For instance, the molecular machine that builds proteins in E. Coli (the ribosome) can be taken apart into 58 isolated protein chains (as well as a few RNA molecules. If these components are mixed back together with the appropriate timing then functional ribosomes can be reassembled. Membrane bound proteins are incorporated into the cellular membranes that surround cells and their compartments. In order for these proteins to perform their numerous functions (signal transduction, ion balance maintenance, molecular transport, electron transport, structure, etc.), they must be appropriately oriented and organized within the membrane. Determining their structure within the membrane is not an easy task for biochemists, since the membranes are often destroyed during purification. However, membrane bound proteins often have a common structural characteristic that suggests a mechanism for their assembly. That is, several hydrophobic residues exposed on a certain area of the protein. Since the center of the membrane is hydrophobic, it can be expected that these hydrophobic residues stick to the inside of the membrane.

Again, the hydrophobic effect is most effective for this purpose. Finally, once the molecule is incorporated into the membrane, changes in the membranes constitution as well as interactions with neighboring membrane bound proteins govern the organization of these proteins. Cellular membranes define the boundries between various cellular compartments (such as the nucleus, mitochondria, endoplasmic reticulum, etc.) as well as enclosing the cell from the external environment. Similar to soap bubbles, cellular membranes are made of specific types of surfactants that tend to form bilayers. The primary surfactants are phospholipids, steroids (primarily cholesterol), and glycolipids. Two nanoscale bodies of water can be effectively separated from one by the formation of a bilayer of these molecules. The main barrier is the hydrophobic region in the center of a lipid bilayer membrane. Water and polar molecules dissolved in water can not pass through such a nonpolar barrier without the assistance of ion channels or molecular pumps.

Development of Molecular Diagnostic Tools

Since the beginning of the 1990's Thailand has led the world in the export of farmed shrimp. In 1999, there were 25,000 shrimp farms producing 240,000 tons of exported black tiger shrimp at a value of 87,000 million baht and employing approximately 130,000 people. Thus, cultivated shrimp is one of the top 10 exports from Thailand, generating approximately 1.5 billion dollars in yearly export earnings. These earnings are especially important because they are based very largely on inputs of local origin. Modern biotechnology has a major role to play in the shrimp cultivation through breeding, feed development and supplementation, disease control, quality assurance and waste treatment and recovery.

The important milestones in the biotechnology application has been the development of DNA probes for the rapid detection of major shrimp pathogens [Flegel, T. W. 1997. Special topic review : major viral diseases of the black tiger prawn [Penaeus monodon] in Thailand. World J. Microbiol. Biotech., 13:433-442.] These probes are essential for the development and monitoring of certified shrimp broodstock and fry. They are also critical tools for the prevention of shrimp diseases which may cause disastrous production losses, even though they present no health risk to shrimp consumers. The development of a DNA probe for one of these pathogens, white spot syndrome virus [WSSV], has yielded a benefit that can be valued at approximately 1 billion US$ per year since 1996. WSSV probably originated in China in 1993 and gradually spread from there to the rest of Asia in succeeding years. In China, it resulted in a drop of shrimp production from approximately 155,000 metric tons in 1992 to 35,000 metric tons [77% decrease] by 1994. [Rosenberry, Bob. 1994. World shrimp farming 1994. Shrimp News International, San Diego.] During that time [1994-1995], the virus had not yet caused any known farm losses in Thailand. In the meantime, work was done at Mahidol University, Thailand to investigate the nature of the virus and to develop diagnostic probes for it. [Wongteerasupaya, C., Wongwisansri, S., Boonsaeng, V., Panyim, P., Pratanpipat, P., Nash, G.L., Withyachumnarnkul, B. and Flegel, T.W. 1996. DNA fragment of Penaeus monodon baculovirus PmNOBII gives positive in situ hybridization with viral infections in 6 penaeid shrimp species. Aquaculture 143: 23-32].

A major discovery with the probe was that shrimp fry used to stock shrimp ponds could be the source of the virus. WSSV began to cause Thai farm losses in late 1995. The research allowed for the development of effective prevention programs which included the use of DNA technology to screen stocking fry so that WSSV positive batches could be rejected. As such, a loss of 60,000 metric tons or 27% of production from the preceding year was recorded. Without research results, the losses would have reached 170,000 metric tons [i.e., 77% of the previous year's production, like China]. The research allowed for the development of effective prevention programs which included the use of DNA technology to screen stocking fry so that WSSV positive batches could be rejected. The loss from WSSV is international. Equador shrimp production dropped from 144,000 tons in 1998 to 95,000 and 45,000 metric tons in 1999 and 2000, respectively. Due to an out brake of WSSV in South America, which is a main shrimp exporter to

U.S.A., the shrimp production in Thailand rose to 280,000 metric tons in the year 2000 compared to 243,000 and 230,000 tons in 1998 and 1999, respectively. The example given here serves to illustrate the research invested can yield substantial returns.

Biosafety of GMOs in Thailand

Thailand is one of the few countries in the world that is not a signing member of the Rio de Janeiro Agreement on Convention of Biological Diversity. However, Thailand has adopted the Biosafety guidelines in 1992 for laboratory work and field work and planned release. The Biosafety protocal was initiated by National Center for Genetic Engineering and Biotechnology (BIOTEC) and the completion was largely the efforts of relevant government agencies. The chronology is as appeared in Appendix 1. In 1990, a biosafety subcommittee was established largely to oversee the drafting of the guidelines for laboratory and field trials of genetically modified plants.

Thailand's first biosafety guidelines were completed in June of 1992. Subsequently in 1993, the National Biosafety Committee (NBC) was established with BIOTEC served as the coordinating body and secretariat. At the same time, Institutional Biosafety Committees (IBCs') were established at various major research and academic institutes throughout Thailand. Currently, there are 14 IBC's including 1 private enterprise laboratory overseeing all the research activities involved the use of genetically modified organisms. [Napompeth B. 1997. Status of Biosafety in Thailand and Development of Biosafety: International Level. A paper presented at the Asia-Pacific Workshop on Biosafety: Environmental Impact Analysis of Transgenic Plants. Madras (Chennai), India, July 21-26, 1997.] Perhaps, the first request of introduction of genetically modified organism in Thailand was the field trial of FLAVR SAVR tomato.

The Department of Agriculture (DoA) of the Ministry of Agriculture and Cooperatives acted with the technical recommendation of NBC, granted permission for the field trial of FLAVR SAVR tomato in 1994. The purpose of FLAVR SAVR tomato field trial was for seed production destined for export only. The request for field trial of genetically modified cotton with toxin gene from Bacillus thuringiensis was made in 1995. The field trial of Bt cotton started in March of 1996 and until today, permission for the commercial release of Bt cotton is still pending by the DoA. In October 1999, with the controversy of the safety of GMOs escalated particularly in the European Union countries, Thailand's Committee for International Economic Policy issued a guideline for the commercial release of seeds derived from genetically modified plants.

The national guidelines prohibited any commercial release of plant seeds. However, field trials are permissible under the jurisdiction of DoA. The 1964 Plant Quarantine Act and the 1999 amendments have restricted importation of over 70 different varieties of transgenic plants. Moreover, products derived from genetically modified plants must be proven as safe before they can be permitted to use as foods or food ingredients. Food safety is to be under the jurisdiction of the Food and Drug Administration (FDA). The Committee for International Economic Policy also agreed to exempt the import restriction of transgenic soybean and maize [Notification of Department of Agriculture, Ministry of Agriculture and Cooperatives] As the technical support to various government agencies in the decision making of the safety of genetically modified organisms, NBC established 3 specialized subcommittees on plant, microorganisms and food.

These subcommittees are to be the technical experts and to work coordinately with relevant government agencies in the approval process. The subcommittee for Food drafted a guideline for safety assessment of genetically modified foods in 1999. It is now being considered by the Thai's FDA for use as a national guideline. The guidelines followed the internationally accepted concept of substantial equivalence. The first transgenic food plant product that is seeking approval for use in food industry is the Bt cottonseed oil. The decision is still pending by the Thai's FDA. Currently, the products being assessed by the Subcommittee includes genetically engineered papaya for viral resistance developed by BIOTEC and DoA and both Bt and Round up Ready maize varieties from Monsanto.

Public Awareness and Information

From the 1970s when genetic engineering just began, we saw the modification of over 60 plant species and nearly 3000 field tests of GM-crops. Now across the world, thousands of varieties are being tested. But while genetically modified organisms have moved rapidly from laboratories to big business, many scientific and technological questions remain. At the same time, public fears over safety and lack of trust in government and private sector management and regulation of the new technology have led to widespread concern over GM-food. Anyhow, biotechnology and gene technology are increasingly becoming part of our world.

It is important that people in our community are made aware of the nature of these technologies and the issues surrounding them. The single biggest problem is public confidence in science. As such, public awareness program is needed to inform the public of biotechnology issues, including benefits, risk factors and the measures used to manage these. It is necessary to consider ways to effectively develop and deliver information to the public. Several booklets on biotechnology and its application. Including GMOs, cloning and world of microbes were published by

BIOTEC and distributed to schools across the nation. A BIOTEC Website [http://www.biotec.or.th] on biotechnology provides access to relevant document and information on latest developments in biotechnology. BIOTEC has also established an electronic forum on GMOs [http://policy.biotec.or.th] for questions and answers and open discussion. A wide-ranging public dialogue on hot issues could be pursued by holding meetings in public and posting minutes to the web. This is one way the government agencies could engage with the public in the decision-making process. But although everyone has bought into this idea, it will take quite a while to ensure that this culture change takes hold.

Genetic Engineering

Genetic engineering (GE) is the manipulation of genetic material (ie, DNA or genes) in a cell or an organism in order to produce desired characteristics and to eliminate unwanted ones. GE includes a range of different techniques with many different uses, and can be applied to plants, animals and humans. For example, the genetic modification of food is a form of GE that involves manipulating the cells of plants such as maize, to increase the yields, make it more nutritious and to make it drought- and disease-resistant. However, the most contentious type of GE is definitely related to its applications in humans. GE in humans has opened up a Pandora's box of possibilities as it can be used for both the miraculous and the sinister. The cloning of Dolly the sheep in 1996 was a very important event. Until then, the cloning of a human was only possible in theory. Recent films and TV programmes such as *Gattaca, Mutant X, Dark Angel* and books such as *Brave New World* all focus on possible consequences of this technology – but what is the real deal?

Cloning

If you were told that there was a clone of you sitting in the next room, what would you expect the clone to look and act like? Probably, exactly the same as you. But, despite all the movies and TV programmes that have explored the possibility of exact clones, it is highly unlikely that a clone of you would look exactly the same and would certainly not act exactly the same. A clone is not completely genetically identical, as there are small differences in the genetic make-up just as there are with identical twins. Despite the fact that identical twins come from the same egg, after a while one begins to notice the differences between them in order to tell them apart. It has been discovered – by doing a number of studies on identical twins brought up in the same environment – that genes mysteriously react differently to the same environment. While cloning a whole human is certainly the ultimate challenge for genetic engineers, cloning is not limited to this goal. Cloning can be, and is, done on a much smaller scale and could involve no more than just the cloning of a single cell. Due to the fact that genetic differences are likely to exist between a clone and its donor, this uncertainty has led to many countries banning the reproductive cloning of humans.

Reproductive cloning

In order to make a clone of someone, one needs a living cell and a human egg (ovum). The nucleus of the egg, which contains the DNA, is removed and replaced with the nucleus from the cell of the person/animal to be cloned. A short electrical pulse then stimulates the egg to start dividing and the embryo is then implanted into the womb where it develops into a duplicate of the person that donated the cell nucleus. Clones created in this way are not 100% genetically identical, as there is some DNA from the original egg cell that is found outside the nucleus (mitochondrial DNA).

Therapeutic cloning and stem cells

In therapeutic cloning an embryo is created in the same way as reproductive cloning, but it is not implanted into the womb of a woman. Instead, *stem cells* are extracted after the embryo starts dividing in the first 14 days after fertilisation, which kills the embryo. Stem cells are special cells with the ability to reproduce and become one of 300 types of cells, eg, skin, liver cell, hair or blood cells. These cells are then used to grow the specific type of tissue or organ that is needed and has the advantage of being genetically identical to the patient who donated it, eliminating the problem of organ or tissue rejection. Currently, if someone has an organ transplant, there is quite a high possibility that their body will reject the foreign organ and so they not necessary have to suppress the immune system to lessen the chances of this happening. Stem cells could potentially be used to repair damaged or defective tissues around the body, such as the cells in the pancreas that stop producing insulin in diabetics.

Adult stem cells versus embryonic stem cells

The problem with embryonic stem cells is that many people feel that by using a human embryo and then killing it, you are actually killing a potential person. It is for this reason that the United States and other countries are calling for a global ban on all human cloning, including the use of embryonic stem cells. As a result, there has been new research into the potential of what is known as "adult stem cells". More than 30 years ago it was discovered that, in addition to being found in embryos, stem cells are also found in adults. Adult stem cells can be removed from a person without causing any harm. Until very recently, it was thought that adult stem cells were only suitable for cloning a few types of cells and tissues. However, new studies have found adult stem cells with almost the same abilities as embryonic stem cells in various human tissues, including: the spinal cord, the brain, connective tissue and in the blood of the umbilical cord. These studies have shown that it is possible for us to manipulate cells to take on new functions/start doing different jobs simply by placing them in different environments and "reprogramming" them. For example, neural (brain) stem cells in mice have been transformed into the blood stem cells that produce the different types of blood cells. In other animals, bone marrow stem cells have become brain cells and liver cells. In humans, bone marrow stems cells of a patient have been transplanted into the areas of the heart damaged by a heart attack, triggering new growth of heart tissue. Although adult stem cells are already being used successfully to treat a number of diseases, there arc a number of problems to be overcome, such as finding and extracting the adult stem cells. Embryonic stem cells have not yet been used for therapy and research is still in the initial stages. Eventually, it may be that, rather than just one type, both types of stem cells will be used for different purposes – whichever proves to be the best for the situation. In the meantime, parallel research continues on both cell types.

Uses of Cloning

Although currently the risks of cloning outweigh the possible benefits, there are many different potential uses of human cloning technology:

- *Replacing organs and other tissues* – such as new skin for burn victims, brains cells for those with brain damage, spinal rod cells for the paralysed and complete new organs (hearts, liver, kidney and lungs). Pigs are also being genetically modified to make their organs more compatible with humans by removing the gene that causes rejection. People could have their appearance changed (cosmetic surgery) using their own cloned tissue and accident victims and amputees could also benefit from this tissue regeneration.

- *Infertility* – human cloning provides couples and individuals who are unable to have children with another potential option.
- *Replacement of a lost child* – parents who have lost a child through an accident or an illness could clone an identical "replacement" child.
- *Creating "donor" people* – cloned people could be created to provide a source of transplant material.
- *Gene therapy* – cloning technology could be used to prevent, treat and cure genetic disorders by changing the expression of a person's genes. This technology may also provide the cure for cancer by revealing how cells are switched on and off. Gene therapy could be used to treat somatic (body) cells where the change is not passed on to children, or germ (egg and sperm) cells where the changes are passed on.
- *Saving endangered species* – by boosting their numbers through creating clones. However, since clones are almost genetically identical, the genetic diversity of the species would not be increased.
- *Reversal of the ageing process* – once more is understood about the role that our genes play in the ageing process. However, some of the above uses carry with them some serious ethical implications.

Problems with Cloning Techniques

Dolly the sheep was created in 1996 using the cloning methods outlined above. Although Dolly was born looking normal, she has suffered from several problems associated with the cloning technique, including premature arthritis, which is thought to be a side-effect of the cloning. Other problems with the current cloning techniques, include:

Allowing Cloning: Around the world, different countries have different rules relating to cloning – some don't allow reproductive cloning, but do allow therapeutic cloning, while other countries allow both types. In South Africa currently, the Human Tissue Act prohibits both therapeutic and reproductive cloning of humans. This situation will change when chapters 6 and 8 of the National Health Act are promulgated - and therapeutic cloning will then be allowed under strict conditions, requiring Ministerial permission, but reproductive cloning on humans will remain strictly prohibited. In China, there are no laws against cloning either. However, with the huge population problem and the policy of only one child per family, there is no interest in reproductive cloning. On the other hand, there is huge interest in therapeutic cloning and, as a result, there is a lot of it on the go in China. Some Chinese cultures believe that humans only become people when they participate in society – so, according to these cultures, embryos and foetuses are not considered to be human beings, thereby eliminating any ethical problems surrounding the creation and destruction of embryos to get the required stem cells. The Chinese government is funding extensive research and drawing back many Chinese scientists from overseas, to undertake work they would not be allowed to do elsewhere. With easy access and no limits on obtaining the embryonic material they require, it is expected that Chinese scientists will race ahead of the rest of the world in therapeutic cloning technologies. In the UK there are clear rules banning reproductive cloning, but scientists in both the UK and Israel are allowed to generate new embryonic cell lines for therapeutic research. The law in Germany bans the extraction of stem cells from human embryos for research within the country, but in 2002 a new law was passed allowing some human embryonic stem cells to be imported. This means that German scientists are allowed to undertake work on embryonic stem cells if they originate from outside Germany. In the United States there is no public funding for research on embryonic stem cells. Therefore, although there is no law against therapeutic cloning in the US, there is almost no public research happening due to the lack of money and access to the embryonic material. Simultaneously, an announcement was recently made stating that US public funds (about US$1,4 million, about R11,2 million) will be provided for studies using adult stem cells instead of embryonic stem cells. These different rules for different countries mean that if a scientist is banned from undertaking cloning in his/her home country, he/she can simply move to a country where it is allowed. To prevent this from happening, the US and about 30 other countries want a global ban on all forms of cloning. However, many countries don't agree and thus no progress is being made on implementing this ban.

Genetic engineering, genetic modification (GM), and the now-deprecated gene splicing are terms for the process of manipulating genes, usually outside the organism's normal reproductive process. It often involves the isolation, manipulation and reintroduction of DNA into cells or model organisms, usually to express a protein. The aim is to introduce new characteristics such as making a crop resistant to an herbicide, introducing a novel trait, or producing a new protein or enzyme. Examples include the production of human insulin through the use of modified bacteria, the production of erythropoietin in Chinese Hamster Ovary cells, and the production of new types of experimental mice such as the OncoMouse (cancer mouse) for research, through genetic redesign. Since a protein is specified by a segment of DNA called a gene, future versions of that protein can be modified by changing the gene's underlying DNA. One way to do this is to isolate the piece of DNA containing the gene, precisely cut the gene out, and then reintroduce (splice) the gene into a different DNA segment. Daniel Nathans and Hamilton Smith received the 1978 Nobel Prize in physiology or medicine for their isolation of restriction endonucleases, which are able to cut DNA at specific sites. Together with ligase, which can join fragments of DNA together, restriction enzymes formed the initial basis of recombinant DNA technology.

Genetic engineering, the use of various methods to manipulate the DNA (genetic material) of cells to change hereditary traits or produce biological products. The techniques include the use of hybridomas (hybrids of rapidly multiplying cancer cells and of cells that make a desired antibody) to make monoclonal antibodies monoclonal antibody, an antibody that is mass produced in the laboratory from a single clone and that recognizes only one antigen. Monoclonal antibodies are typically made by fusing a normally short-lived, antibody-producing B cell (see immunity) to a fast-growing cell, such as a cancer cell (sometimes referred to as an "immortal" cell).; gene splicing or recombinant DNA, in which the DNA of a desired gene is inserted into the DNA of a bacterium, which then reproduces itself, yielding more of the desired gene; and polymerase chain reaction polymerase chain reaction (pOl`-mYrâs') (PCR), laboratory process in which a particular DNA segment from a mixture of DNA chains is rapidly replicated, producing a large, readily analyzed sample of a piece of DNA; the process is sometimes called DNA amplification, which makes perfect copies of DNA fragments and is used in DNA fingerprinting

DNA fingerprinting or DNA profiling, any of several similar techniques for analyzing and comparing DNA from separate sources, used especially in law enforcement to identify suspects from hair, blood, semen, or other biological materials found at the scene of a violent crime. Monoclonal antibody, an antibody that is mass produced in the laboratory from a single clone and that recognizes only one antigen. Monoclonal antibodies are typically made by fusing a normally short-lived, antibody-producing B cell (see immunity immunity, ability of an organism to resist disease by identifying and destroying foreign substances or organisms. Although all animals have some immune capabilities, little is known about nonmammalian immunity. Mammals are protected by a variety of preventive mechanisms, some of them nonspecific (e.g., barriers, such as the skin), others highly specific (e.g., the response of antibodies). Nonspecific Defenses) to a fast-growing cell, such as a cancer cell (sometimes referred to as an "immortal" cell). The resulting hybrid cell, or hybridoma, multiplies rapidly, creating a clone that produces large quantities of the antibody.

Monoclonal antibodies engendered much excitement in the medical world and in the financial world in the 1980s, especially as potential cures for cancer. They have been used in laboratory research and in medical tests since the mid-1970s, but their effectiveness in disease treatment has been limited. By the mid-1990s, however, some of the technical problems had been overcome. Experimental cancer therapies have used drugs, radioactive materials, or immune killer cells attached to monoclonal antibodies that, when injected into patients, home in on antigens that grow only on the surface of cancer cells.

Genetically engineered products include bacteria designed to break down oil slicks and industrial waste products, drugs (human and bovine growth hormones, human insulin, interferon), and plants that are resistant to diseases, insects, and herbicides, that yield fruits or vegetables with desired qualities, or that produce toxins that act as pesticides. Genetic engineering techniques have also been used in the direct genetic alteration of livestock and laboratory animals (see pharming pharming (fär'm-ng), the use of genetically altered livestock, such as cows, goats, pigs, and chickens, to produce medically useful products. In pharming, researchers first create hybrid genes using animal DNA and the human or other gene that makes a desired substance, such as a hormone.). Genetically engineered products usually require the approval of at least one U.S. government agency, such as the Dept. of Agriculture, the Food and Drug Administration, or the Environmental Protection Agency.

Because genetic engineering involves techniques used to obtain patents on human genes and to create patentable living organisms, it has raised many legal and ethical issues. The safety of releasing into the environment genetically altered organisms that might disrupt ecosystems has also been questioned. The discovery in 2001 of genetically engineered DNA in native Mexican corn varieties made concerns of genetic pollution actual, and led some scientists to worry that the spread of transgenes through cross-pollination could lead to a reduction in genetic diversity in important crops. Imports of genetically modified corn, soybeans, and other crops have been curtailed or limited in some countries, and the vast majority of such crops are grown in just a handful of nations. The Cartagena Protocol on Biosafety, which has been signed by more than 100 nations and took effect in Sept., 2003, requires detailed information on whether and how imported seeds, plants, animals, other organisms, and the like are genetically modified and permits a nation to bar those imports. The United States, however, is not party to the treaty.

When Hungarian engineer Karl Ereky coined the word biotechnology in 1919 to describe the production of goods from raw materials with the aid of living organisms, he was describing techniques that had been employed by human beings for centuries. This includes techniques such as selective breeding, fermentation, hybridization, phyto-pharmacology, vaccination or the use of biomass for energy production.

However, the term biotechnology has evolved to include modern techniques with new applications, to what is now known as modern biotechnology. Modern biotechnology encompasses techniques such as genetic engineering (which is the direct intervention on the genetic makeup of an organisms usually by introducing foreign DNA into its gene pool by means that would not occur naturally), tissue culture and others. These modern techniques also take considerably less time to achieve the desirable changes in living organisms and carry greater precision than traditional techniques. In short, genetic engineering is a specific subset as a part of the more general subject of biotechnology.

Human genetic engineering refers to the the controlled modification of the human genome. DNA provides the the genetic blueprint for all living organisms and can influence individuals' actions and abilities. With the advent of DNA research and the ability to change gene expressions, it is now possible that scientists may be able to change human capacities, whether they be physical, cognitive, or emotional. Human genetic engineering is still in its infancy, however, with current research generally restricted to animals or gene therapy. Healthy humans do not need gene therapy to survive, though it may prove helpful to treat certain diseases. Special gene modification research has been carried out on groups such as the 'bubble children' - those who's immune system do not protect them from the bacteria and irritants all around them. The first clinical trial of human gene therapy began in 1990, but (as of 2006) gene therapy is still experimental.

Other forms of human genetic engineering are still theoretical, or restricted to fiction stories. Recombinant DNA research is usually performed to study gene expression and various human diseases. Some drastic demonstrations of gene modification have been made with mice and other animals, however: testing on humans is generally considered off-limits. In some instances changes are usually brought about by removing genetic material from one organism, and transferring them into another species. This method is known as recombinant genetics. There are two main types of genetic engineering. Somatic modifications involve adding genes to cells other than egg or sperm cells. For example. if a person had a disease caused by a defective gene, a healthy gene could be added to the affected cells to treat the disorder. The distinguishing characteristic of somatic engineering is that it is non-inheritable, e.g. the new gene would not be passed to the recipient's offspring. Germline engineering would change genes in eggs, sperm, or very early embryos. This type of engineering is inheritable, meaning that the modified genes would appear not only in any children that resulted from the procedure, but in all succeeding generations. This application is by far the more consequential as it could open the door to the perpetual and irreversible alteration of the human species. Genetic engineering is most easily accomplished by making changes just after the egg and sperm have melded but before first division. In this way, the gene will be expressed throughout and will affect the recipients children, grandchildren, and all subsequent generations. This type of germline engineering is highly controversial and deemed inappropriate by most scientists. As of now, this is likely to take the form of gene therapy. This is less controversial as the engineered person can give consent, unlike germline engineering where subsequent generations are brought into the world with a genetic modification made to their body without their consent. Each human cell contains a different set of genes, so different genes could be transferred to different parts of the body. Such changes will not be hereditary unless the sex cells are engineered.

Biotechnology Ethics

The explosion of genetic knowledge comes with some heavy ethical and social baggage. It is not clear how society will deal with the growing potential to manipulate genomes and ethical issues bring on dilemma which could not be clearly defined. The U.S. government is grapping with how to protect individual rights once the technology exists for reading each person's genome. Meanwhile, it established an Office for Human Research Protections within the Department of Health and Human Services. The U.S. government spent about 5% of its human genome project to study the ethical, social, and legal impacts of genomic research.

The German government plans to use US$ 10 million on this issue. Meanwhile, there are attempts from international organizations to set up guidclines for ethical committee. UNESCO set up International Bioethics Committee (IBC) to set standards and regulations. World Medical Association announced through Helsinki Declaration on Ethical Principles for Medical Research Involving Human Subjects. WHO proposed International Guidelines on Ethical Issues in Medical Genetics and Genetics Services. All of which are prerequisites for newly developed biotechnology-related guidelines which will potentially be established in the near future.

In Thailand, some guidelines, particularly in genetics,

will be initiated directly by BIOTEC, in co-operation with WHO and UNESCO. For Thailand, ethical considerations regarding biotechnology are becoming focus, particularly on advanced biomedical research area. BIOTEC and National Health Foundation are joining hands on the special project namely "Bioethics and Advanced Biomedical Research". As some advanced technologies bring along new frontier confrontation of ethical issues, therefore, specific technologies are chosen as high-impact ones. These are gene therapy, gene hunting, stem cell research, genetic testing, human genome database, human cloning and intellectual property on human genetics. Problems and dilemmas are provoked upon such technologies regarding discrimination, human rights, eugenics, and rights on one's own genetic materials. BIOTEC, therefore, initiated Bioethics Advisory Committee in order to consider, analyse and make recommendations on such special issues.

The committee comprises experts from science, research, medical, religion, law and human rights sectors. Web site (www.policy.biotec.or.th) and InsightBio newsletter are actively created, updated, produced and widely distributed among potential stakeholders in order to pass on information and create awareness throughout Thailand. Moreover, in-dept research and seminars on ethical issues are continually supported. Public understanding, both directs to public and pass through news agents, is considered as a significant process to stimulate public understanding. Above all, the reasonable and acceptable policy on bioethics will be strongly initiated and clearly issued for the sake of Thai society.

Biotechnology and Intellectual Property Rights

The field of intellectual property protection for biotechnology is somewhat new, as is the biological technology itself. The current impulse that influenced the expansion of intellectual property protection to biotechnology was the advancement of recombinant DNA technology in 1970's. Nowadays, biotechnology is relatively important in multiple practices. Consequently, sufficient intellectual property rights for new inventions and other progresses in biotechnology have been called for by the developed countries who dominate this field. Thus far, many developed countries tried to embody the recognition of biotechnology in various intellectual property regimes, such as the Agreement on Trade Related Aspects of Intellectual Property Rights (TRIPS), an agreement under the umbrella of the World Trade Organization (WTO). Many forms of intellectual property rights, e.g., patent, protection of trade secret, and plant variety protection, may be applicable to biotechnology. With respect to patent protection, developed countries, especially the U.S. tried to include protections for every kind of living thing, beginning with microorganisms, in the framework of the WTO, specifically in TRIPS.

Developed countries went further, asserting that the existing intellectual property regimes in most countries should be reformed to include the protection of biotechnology, and that the new intellectual property rights should be extended to developing countries like Thailand as well in order that the rights granted will be recognized worldwide. The law of trade secret may also be a practicable form of protection for particular forms of biotechnology. When the biotechnology invention cannot be effortlessly reverse engineered the law of trade secret may be preferable to patent. Trade secret may also be an alternative, if patent protection is not available for an invention.

Today, research and development in biotechnology is believed to be enhanced with the support of intellectual property rights. New techniques for improving cells promise to aid in the production of useful substances, whereas intellectual property rights vow to confer the ultimate benefit on inventors of the said techniques. In the biotechnology industry of many industrialized countries, scientists use genetic manipulation techniques to modify animal and plant's genes and cells to find new hybridizes which are more productive than the parental. These technologies would succeed in a country like Thailand if they are used with conventional breeding by providing agricultural services, incentive price policies for agricultural products, and effective marketing network. With respect to intellectual property protection, Thailand has enacted several laws that are applicable for the protection of biotechnogical products and processes, e.g., the Patent Act, the Plant Variety Protection Act, the Bill on the Law of Trade Secret (draft of the Trade Secret Act), etc.

The most used and relevant protection for biotechnology industry is, however, the law of patent similar to any other existing patent laws worldwide, Thai patents are given to inventions and processes, which are new, possess an inventive step, and are capable of industrial application. The protection is given to both product and process of biotechnology. Notwithstanding, patent eligibility is not given to naturally existing microorganisms and their components, animals, plants or animal and plant extracts. Ironically, the law always finds a hard time coping up with technologies. Intellectual property law is no exception to such matter of fact.

As may be seen from this paper that the advancement of biotechnology in Thailand has moved further into gene technologies, and to genomic and proteiomic inventions: nevertheless the Thai Patent Act is still struggling with protection for DNA, genes, and proteins, especially those extracted, isolated, and purified from plants or animals. This is because the products of which are considered animal or plant extracts that hold no patentability under the Thai law. The issue of patentable subject matters is indeed not only the question of law but also the consideration of policy. Patent protection as well as any other intellectual property rights could be one of the promising tools to confer incentive on biotechnological research and development.

Therefore, Thailand should think through carefully, especially in the post-genomics era, as to what role would intellectual property rights be playing in order to enhance the progression of the country's biotechnology industry so as to scale up the overall economic development and the betterment of science and technology sector as a whole. Initially, Thailand should not focus on competing with industrialized countries in sophisticated biotechnologies, such as gene transfer and genetic engineering. The country should take advantages of simple biotechnology, for instance, simple technologies of plant tissue, in order to achieve swift vegetative propagation of valuable crops.

There should be an adoption and improvement of specific plant biotechnology methods that are less expensive, easy to establish, and appropriate to transfer and adapt to local situations. Moreover, in avoid using sophisticated technologies developed by industrialized countries, Thailand would be immuned from the intellectual property right infringement: and, in developing her own technologies, the country would possess intellectual property rights of her own. This suggestion does not mean that Thailand will not have access to sophisticated biotechnologies.

In fact, biotechnologies vary in complexity and sophistication; therefore, they should be distributed to communities by the ability of development techniques. The national scientific or technological community should be responsible for sophisticated technologies then adapt them to development projects to serve local needs and keep abreast of new trends in biotechnology. Thailand needs to determine her priorities by identifying economic objectives and considering how to maximize the value of available resources. The government should support research and development in biotechnology by funding national scientific communities and credit loans to the private sectors. The government should ensure the transfer of technology to the local communities in order to enhance techniques of traditional breeding, and increase quality and quantity of agricultural products.

Modern biotechnology is already making important contributions and poses significant challenges to agriculture, health and environment. Biotechnologies are a new group of powerful tools for research and ultimately for accelerating development, and not an end in themselves. Modern biotechnologies should be used as adjuncts to-and not as substitutes for conventional technologies in solving problems and that their application should be need driven rather than technology-driven. Biotechnology is research based and multidisciplinary. Successful development and application of biotechnology are possible only when a broad research and knowledge base in several subjects such as, microbiology, biochemistry, molecular biology and plant breeding exists. Biotechnological programs must be fully integrated into a research background and a continued commitment to basic research is a must to fulfill benefits offered by the emerging technologies. Thailand's strategy is to keep biotechnology in a balanced perspective by undertaking activities within the framework of existing national research agendas and priorities.

Main Biotechnology Applications

Modern biotechnology techniques are currently being used in many areas such as food, agriculture, health care, forestry, the environment, minerals, oil and gas recovery and industrial processes to develop new products and processes, and to modify existing ones. Summarized below are some of the ways that modern biotechnology techniques have been applied in these areas.

Forestry & Biotechnology

To date, only a few biotechnology-derived products and processes used in forestry have been commercialized. Most biotechnology applications in forestry are still in the research and development stage. However, some commercial applications include the development of bacterial bio-pesticides such as *Bacillus thuringiensis* (Bt), as an alternative to chemical pesticides. Tissue culture technology is also being used to produce genetically modified seedlings for forest regeneration. For example, BCRI Inc., a biotechnology company in British Columbia, is using tissue culture to produce weevil-resistant Sitka spruce. Companies are also developing bio-fertilizers such as plant growth-promoting rhizobia bacterial fertilizers. In addition, genetic engineering has been used to develop traits of agronomic interest such as pest and disease resistance, within trees. The pulp and paper industry has used enzymes and microorganisms in their bio-bleaching and de-inking processes.

Environment & Biotechnology

Biotechnology applications in the environment focus on using living organisms primarily to treat waste and prevent pollution. Examples of these applications include bio-filtration and bio-remediation. Bio-filtration refers to the use of microorganisms to remove complex pollutants from the air emissions and waste water discharges of many manufacturing processes. Bio-remediation refers to a number of processes that use living organisms to degrade toxic waste into harmless by-products such as water, carbon dioxide and other materials. Examples of bio- remediation processes include:

- *Bio-stimulation:* A technique which involves introducing nutrients to stimulate the growth of waste-eating microorganisms already present in the environment at a waste site.
- *Bio-augmentation:* A technique whereby microorganisms, not normally present in the ecosystem of a contaminated site, are added to the site to clean up pollutants. The microorganisms added may be natural or genetically modified. The organisms die when their food is used up.
- *Phytoremediation:* A technique which uses certain plants and fungi, planted at waste sites, to naturally reclaim the sites without using chemical treatment, incineration or land filling. These plants and fungi are used because they flourish by accumulating metals and other waste materials present in the soil.

Industrial Processes & Biotechnology

Biotechnology has been applied in a variety of industrial processes in different ways, particularly in the use of biocatalysts in manufacturing processes. Biocatalysts are substances that initiate or modify the rate of a biological process and are generally consumed in the process. Some examples of industrial processes where biotechnology has been applied include chemicals, starch/grain processing, cleaning and textile industrial processes. In the chemicals industry, biotechnology has been used to produce commodity and specialty chemicals. It is also used in the starch and grain processing industries through the use of enzymes to turn starch into glucose and fructose. Corn and other grains can be converted to sweeteners such as high-fructose corn syrup and maltose syrup, using enzymes. Biotechnology applications have been used to produce ethanol from grain. Furthermore, biotechnology is used in the textile industry for the finishing of fabrics and garments. In the pulp and paper industry, bio-pulping is used to manufacture some products.

Energy & Biotechnology

To date, biotechnology applications in the energy industry include the production of cleaner coal and petroleum by removing sulfur and thus reducing the environmental contaminants released during combustion. Biotechnology has also been used widely to produce bio fuels such as bio-ethanol and bio-diesel. Bio fuels are alcohols, ethers, esters and other organic chemicals made from biomass such as herbaceous and woody plants, agricultural and forestry residues and industrial waste. These fuels can be used for electricity and as fuels for transportation.

Food, Agriculture & Biotechnology

One of the most extensive applications of biotechnology has been in agriculture. Biotechnology techniques such as recombinant DNA techniques and mutagenesis have been used to develop plants with novel traits. These traits include herbicide tolerance and pest, insect and virus resistance. Biotechnology techniques have been used to produce biopesticides which are toxic to targeted plant pests. Also, experiments into genetic modification of aquatic organisms such as salmon, for such novel traits as enhanced growth, have been carried out using recombinant DNA techniques.

Health Care & Biotechnology

To date, applications of biotechnology in health care have focused on fighting disease using the human body's own 'weapons'. Biotechnology medicines and therapies synthesize proteins, enzymes, antibodies and other substances which occur naturally in the human body, to fight infections and diseases. However, biotechnology also uses other living organisms, that is, plant and animal cells, viruses and yeasts to help produce human medicines. There are four main areas in health care in which biotechnology is currently being used: medicines, vaccines, diagnostics and gene therapy. Several biotechnology medicines have been developed and approved to treat such diseases as anaemia, growth deficiency in children, hemophilia, diabetes and others. Biotechnology companies have also developed therapeutic products for infectious agents such as HIV, hepatitis B and influenza. Biotechnology is responsible for new vaccines such as the hepatitis B vaccine which was approved in the U.S. to fight the hepatitis B virus. Biotechnology diagnostics have been used to detect a wide variety of diseases and genetic conditions. For example, gene therapy is used diagnostically to screen donated blood to protect the blood supply from HIV and hepatitis. Biotechnology diagnostics are also used in home pregnancy tests. In a home pregnancy test kit, a protein called a monoclonal antibody (MAb), binds to Human Chorionic Gonadotrophin (HCG), causing a colour change. HCG is present in a woman's urine only during pregnancy. In gene therapy, genes are used as 'drugs' to treat hereditary genetic disorders. Faulty or missing genes can be replaced to prevent the occurrence of a genetic disease. Current gene therapy is primarily experiment-based, with a few human clinical trials, such as for the treatment of cystic fibrosis, in the early stages.

Metals, Minerals Recovery & Biotechnology

Biotechnology has been used in the mining industry in such processes as bio-leaching and metal bio-remediation and recovery. Bio-leaching is the use of bacteria to extract valuable metals. Metal bio-remediation and recovery refers to the use of biologically created enzymes in the degreasing process to recover metals of value.

Human Applications & Biotechnology

When treating problems that arise from genetic disorder, one solution is gene therapy. A genetic disorder is a situation where some genes are missing or faulty. When this happens, genes may be expressed in unfavorable ways or not at all, and this generally leads to further complications. The idea of gene therapy is that a non-pathogenic virus or other delivery system can be used to insert a piece of DNA—a good copy of the gene—into cells of the living individual. The modified cells would divide as normal and each division would produce cells that express the desired trait. The result would be that he/she would then have the ability to express the trait that was previously absent at least partially. This form of genetic engineering could help alleviate many problems, such as diabetes, cystic fibrosis, or other genetic diseases.

The potential of genetic engineering to cure medical conditions opens the question of exactly what such a condition is. Some advocates see aging and death as medical conditions and engineering problems to be solved. They see human genetic engineering as a key tool in this *(see life extension)*. The difference between cure and enhancement from their perspective is merely one of degree. Theoretically genetic engineering could be used to drastically change people's genomes which could enable people to regrow limbs, the spine, the brain. It could also be used to make people stronger, faster, smarter, or to increase the capacity of the lungs, among other things. If a gene exists in nature, it could be brought over to a human cell. Others feel that there is an important distinction between using genetic technologies to treat those who are suffering and to make those who are already healthy superior to others. There is widespread agreement that germline engineering should not currently be allowed for either therapeutic and enhancement applications, as evidenced by a recent report by the American Association for the Advancement of Science. Though theory and speculation suggest that genetic engineering could be used to make people stronger, faster, smarter, or to increase lung capacity, the AAAS report finds that there is little evidence that this currently can be done safely or without unethical human experiments.

"Transgenic organism" is now the preferred term for genetically modified organisms with extra-genome (foreign genetic) information, as opposed to "genetically engineered" or "genetically modified" organisms (which may refer to changes made within the genome such as amplification or deletion of genes). The first Genetically Engineered drug was human insulin approved by the USA's FDA in 1982. Another early application of GE was to create human growth hormone as replacement for a drug that was previously extracted from human cadavers. In 1986 the FDA approved the first genetically engineered vaccine for humans, for hepatitis B. Since these early uses of the technology in medicine the use of the GE has expanded to supply many drugs and vaccines. One of the best known applications of genetic engineering is that of the creation of genetically modified organisms (GMOs).

There are potentially momentous biotechnological applications of GM, for example oral vaccines produced naturally in fruit, at very low cost. A radical ambition of some groups is human enhancement via genetics, eventually by molecular engineering. DNA sequencing is a technique which is used to identify each base in DNA. Although the costs of DNA sequencing has dropped dramatically, the NIH estimates it costs at least $10 million to sequence 3 billion base pairs - the size of the whole human genome. Although there has been a tremendous revolution in the biological sciences in the past twenty years, there is still a great deal that remains to be discovered. The completion of the sequencing of the human genome, as well as the genomes of most agriculturally and scientifically important plants and animals, has increased the possibilities of genetic research immeasurably. Expedient and inexpensive access to comprehensive genetic data has become a reality, with billions of sequenced nucleotides already online and annotated. Now that the rapid sequencing of arbitrarily large genomes has become a simple, if not trivial affair, a much greater challenge will be elucidating function of the extraordinarily complex web of interacting proteins, dubbed the proteome, that constitutes and powers all living things. Genetic engineering has become the gold standard in protein research, and major research progress has been made using a wide variety of techniques, including:

- Loss of function, such as in a knockout experiment, in which an organism is engineered to lack the activity of one or more genes. This allows the experimenter to analyze the defects caused by this mutation, and can be considerably useful in unearthing the function of a gene. It is used especially frequently in developmental biology. A knockout experiment involves the creation and manipulation of a DNA construct in vitro, which, in a simple knockout, consists of a copy of the desired gene which has been slightly altered such as to cripple its function. The construct is then taken up by embryonic stem cells, where the engineered copy of the gene replaces the organism's own gene. These stem cells are injected into blastocysts, which are implanted into surrogate mothers. Another method, useful in organisms such as Drosophila (fruit fly), is to induce mutations in a large population and then screen the progeny for the desired mutation. A similar process can be used in both plants and prokaryotes.

- 'Tracking' experiments, which seek to gain information about the localization and interaction of the desired protein. One way to do this is to replace the wild-type gene with a 'fusion' gene, which is a juxtaposition of the wild-type gene with a reporting element such as Green Fluorescent Protein (GFP) that will allow easy visualization of the products of the genetic modification. While this is a useful technique, the manipulation can destroy the function of the gene, creating secondary effects and possibly calling into question the results of the experiment. More sophisticated techniques are now in development that can track protein products without mitigating their function, such as the addition of small sequences which will serve as binding motifs to monoclonal antibodies.
- Gain of function experiments, the logical counterpart of knockouts. These are sometimes performed in conjunction with knockout experiments to more finely establish the function of the desired gene. The process is much the same as that in knockout engineering, except that the construct is designed to increase the function of the gene, usually by providing extra copies of the gene or attracting more frequent transcription.

3

Future Prospects of Biotechnology Science

In the future, biotechnology is used for producing existing products in new ways, identifying new product opportunities (as in drug discovery), and for producing new products that could not be commercially produced before (as with many large molecule therapeutics and some GM plant varieties). The wide range of uses for biotechnology means that it is a generic technology with applications in many different economic sectors. Biotechnology is also better described as a group of related biotechnolog*ies* The diverse types of biotechnologies and the range of possible applications create two main challenges for developing comparable biotechnology statistics: how to define biotechnology and how to define a biotechnology firm

The OECD has developed both a single definition of biotechnology and a list-based definition (see Box below) of different types of biotechnology. The single definition defines biotechnology as "*the application of science and technology to living organisms, as well as parts, products and models thereof, to alter living or nonliving materials for the production of knowledge, goods and services.*" The OECD list-based definition, or close variants, were used in surveys in 15 countries, but different definitions of biotechnology were used in the other 11 countries: 7 studies limit biotechnology to 'modern' or third-generation biotechnologies that are similar to the OECD list-based definition in practice, 2 studies use mixed definitions that include second generation biotechnologies (Japan and South Africa), and 2 do not define biotechnology, but leave it to the survey respondent to decide if their firm is active in biotechnology. As the latter two studies cover Denmark and Sweden, a large majority of the respondents are likely to interpret biotechnology as modern biotechnology

Accuracy of the Results

Three characteristics that could influence the accuracy of the biotechnology statistics: the organisation conducting thc survey (Source), the response rate (RR), and whether or not results were imputed to account for non-respondents, or if a random survey, extrapolated to the total population (Extrapolation) Twenty of the 33 studies were conducted by government agencies (GOV), 6 by non-profit organisations acting at the request of a government agency (NP-GOV), and 7 by a consulting agency. Generally, governments have the most resources to conduct a full survey of biotechnology and consequently the data quality should be better than the results obtained by consulting agencies, although this might not always be the case Data accuracy is strongly dependent on a high response rate, which will reduce possible biases due to differences in the types of biotechnology firms that choose or do not choose to respond to a survey and include most firms active in biotechnology. Low response rates could result in serious underestimates of the number of biotechnology-active firms or biotechnology activities such as R&D

Surveys were not used in six studies, so the response rate is not relevant. Out of the 27 surveys, 14 had high response rates above 80%, while 4 had low response rates (50% or less). When the response rate is not 100%, the results can be extrapolated to estimate biotechnology activity in the estimated number of biotechnology firms that did not respond to the survey. However, the accuracy of extrapolation depends on having good information on the number of eligible non-respondent firms. It is also possible that this number is zero, for example if firms that were not active in biotechnology did not answer the questionnaire because they did not consider it relevant to them

The combination of low response rates and no extrapolation is likely to notably underestimate biotechnology activity by firms in five countries: Belgium, China (Shanghai), Italy, Poland, and the United States (2001 survey only). Moderate underestimating is possible in Japan and South Africa

Cross-country Results for Employment, R&D and Sales

Different definitions of biotechnology activities create problems for comparing employment, R&D and sales across countries. Most consulting reports are limited to total employment and total R&D employment in core biotechnology firms. The advantage of some of the Government-supported surveys is that they cover all biotechnology-active firms and include questions that can differentiate between employees with and without biotechnology-related responsibilities. These types of studies, as in Canada and the United States, provide the most reliable estimates of biotechnology employment. In Canada, total employment among biotechnology-active firms is 75,448 in 2003, but only 11,863 of these employees (15.7%) have biotechnology responsibilities (including biotechnology-related R&D, management, production, and administration). The gap between biotechnologyactive employment and total employment increases with firm size, as larger firms are more likely to have diversified activities. The 2001 American survey finds that less than 20% of employees in biotechnology-active firms with over 2,500 employees have biotechnology-related responsibilities, compared to 60% of employees in biotechnology firms with 51 to 500 employees and approximately 70% of employees in biotechnology firms with less than 50 employees. These differences in study design can result in both over- and underestimates of biotechnology employment. Studies limited to core biotechnology firms will underestimate biotechnology employment by failing to account for biotechnology employment in large firms. This probably explains why the Critical I estimate of total employment in core biotechnology firms in Belgium is only 2,676 employees, while the survey of all biotechnology-active firms estimates total employment of 11,137, of which 4,261 have biotechnology-related responsibilities. Conversely, studies that assume that all employment in biotechnology firms is 'biotechnology' employment can produce overestimates, since an unknown percentage could be involved in activities that do not involve biotechnology

The most comparable estimate between studies that only provide data for total employment in core biotechnology firms, and studies of all biotechnology-active firms, is to use counts of biotechnology-active employees from the latter type of study. For the two countries (Germany and Belgium) where both types of data are available, the number of biotechnology-active employees is greater than total employment in core biotechnology firms. This suggests that using total employment in core biotechnology firms is more likely to underestimate than overestimate total biotechnology employment, although there could be exceptions. Critical I probably overestimates Danish biotechnology employment because it assigns all Danish employment in one or more large pharmaceutical firms to biotechnology. For this reason, some of the Critical I results for Denmark are not included in this report

The same problem as described above for employment also applies to statistics for biotechnology R&D and sales of biotechnology products. Several countries collect data that differentiates between biotechnology R&D and other types of R&D, and between sales of biotechnology products and sales of other products. Studies of core biotechnology firms usually assume that all R&D performed by these firms is biotechnology R&D and all sales are of biotechnology products

The study from Canada shows that not all sales and R&D concern biotechnology. Biotechnology revenues in Canada in 2003 were only 12.5% of total revenues of biotechnology-active firms. The share of biotechnology R&D was higher however, at 65% of total R&D expenditures by these firms

Although every effort has been made to maximise comparability across countries, caution must be used in comparing biotechnology activities between countries when the data are obtained from studies with very different methodologies. This particularly applies to differences between studies of core biotechnology firms only versus studies of all biotechnology-active firms. Other factors, such as differences in the definition of biotechnology, whether or not all firms must innovate, and low response rates in some countries, will also reduce comparability

In general, the results in this document (for all countries combined) for the number of biotechnology firms, employment, R&D and sales are more likely to underestimate the true values than to overestimate them. This is primarily caused by the reliance for several countries on studies limited to core biotechnology firms and the low response rates in several studies

As a final caution, some of the results for specific countries vary, depending on the data source. For example, the 2001 Department of Commerce survey estimates total biotechnology R&D of USD 16,834 million, while the 2003 R&D survey for the United States estimates total biotechnology R&D of USD 14,232 million and is used on page 17. The difference is not likely to be due to the different years but to the different methodologies

The 2003 results are likely to be more accurate, but they are not available by application field. Consequently, the 2001 results are used. Similar discrepancies occur for a few other countries with multiple data sources, including when Critical I data are used in some cases and survey data in other cases. In all cases, the selection of which data source to use is based on the best available data quality, with preference given to survey research with high response rates. We apologise to the reader for any confusion that this may cause

Biotechnology Patents

The following was extracted from the OECD (2005), *Compendium of Patent Statistics*. The report is available on line at: www.oecd.org/sti/ipr-statistics

- Biotechnology patents have grown more rapidly than overall patent applications at the European Patent Office (EPO). Between 1991 and 2002, they grew by 8.3% a year, while total EPO patent applications grew by 5.7%. The rate of increase in biotechnology patents accelerated from 1994 onwards. The latest available data show a slight decline in biotechnology patent filings at the EPO. This could be related to the adoption by the EPO of more restrictive policies for examining biotechnology patents in recent years. This trend is also observed for countries with a large biotechnology patent portfolio

- The latest available data show that around 5.3% of all EPO patent applications are in the biotechnology field (see methodological box for definition). However, the ratio of biotechnology patents to all EPO patents varies substantially across countries. New Zealand, Denmark and Australia have a very high ratio of biotechnology patents to all EPO patents (more than 10%). In contrast, only around 2% of all EPO patents originating from Finland and Italy are biotechnology patents
- For the majority of countries, the ratio of biotechnology patents to all EPO patents increased between 1991 and 2002. However, notable exceptions are Denmark and Belgium as well as countries with a low ratio of biotechnology patents to all EPO patents
- In 2002, more than 5,800 biotechnology patents were filed at the European Patent Office (EPO), most of which originated from the United States (39.9%) and the European Union (34.5%). Around 14% of the EPO biotechnology patents originate from Japan. Since 1997, the shares of the European Union and Japan in biotechnology patents have increased, while those of the United States have continuously decreased
- To measure a country's level of specialisation in biotechnology patents, country shares can be expressed in terms of a specialisation index (see methodological box). By this measure, the United

States is highly specialised in biotechnology patents, while the European Union and Japan are not

- Denmark, Canada, New Zealand and Australia are the most specialised countries in biotechnology patents. Italy, Turkey and Luxembourg are the least specialised. Three non-OECD countries (Israel, China and India) are also relatively specialised in biotechnology patents

Definition of Biotechnology Patents

The definition of biotechnology patents covers the following IPC classes: A01H1/00, A01H4/00, A61K38/00, A61K39/00, A61K48/00, C02F3/34, C07G(11/00, 13/00, 15/00), C07K(4/00, 14/00, 16/00, 17/00, 19/00), C12M, C12N, C12P, C12Q, C12S, G01N27/327, G01N33/(53*, 54*, 55*, 57*, 68, 74, 76, 78, 88, 92)

For further details on the IPC classes, see www.wipo.int/classifications/fulltext/new_ipc/index.htm

The specialisation index (SI) is calculated as the share of country A in a specific technology area (*i.e.* biotechnology patents) divided by the share of country A in all technology areas (total EPO patents of country A). When the SI of biotechnology patents is greater than 1, the country has a higher share in biotechnology patents relative to its share in all technology areas. Conversely, when the SI of biotechnology patents is below 1, the country has a lower share in biotechnology patents than in all technology areas combined

Biotechnology in Agriculture

- Biotechnology has many applications in agriculture, including diagnostics, vaccines and therapeutics for animal health; DNA fingerprinting for managing animal stocks and identifying specific plant varieties, animal and plant propagation, marker-assisted selection for plant and animal breeding programmes, and genetic modification (GM) of plant and animal varieties using rDNA technology. Animal health applications are combined in this report with human health applications, due to the similarity in the underlying biotechnologies

- Internationally comparable data for biotechnology applications in agriculture are limited to GM plant varieties. Two types of data are available: hectares planted with GM crops and field tests of new GM varieties

Hectares planted with GM crops

- GM crops were first extensively commercialised in 1996. Data on GM crop hectares are available for four OECD countries outside the European Union members, plus for India, China, Argentina and Brazil. The number of hectares planted with GM crops is negligible in most European Union countries, although some commercial GM crops are reported in 2005 for Spain, Germany, Portugal, France and the Czech Republic
- In 2005, the United States accounted for 89% of GM crop hectares within the OECD. The United States' share of global GM crop hectares declined from 68% in 2000 to 55% in 2005, as the number of hectares planted with GM crops increased rapidly in other countries
- In the three years between 2002 and 2005, the number of hectares planted with GM crops increased 27% in Argentina, 28% in the United States, 53% in Canada, 57% in China, and 169% in Brazil
- The maximum area that can be planted with GM

crops partly depends on the total arable land under cultivation. Argentina has the highest percentage of arable land planted with GM crops in 2005, at 61%, followed by the United States (29%), Brazil (16%) and Canada (13%). The estimate of 0.5% of arable land in the EU-25 under GM crops is likely to substantially overestimate the actual value

- Two other main factors that determine the area planted with GM crops are regulations and the types of crops grown in each country. GM use is highest for countries with suitable growing conditions for the main GM crops to date: soybeans, maize, cotton, and canola

Arable land is defined as land under annual crops (double-cropped areas are counted only once), temporary meadows for mowing or pasture, land under market and kitchen gardens and land temporarily fallow (less than five years). The abandoned land resulting from shifting cultivation is not included in this category. Data for "Arable land" does not measure the amount of land that is potentially cultivable

Field trials of GM varieties

- For regulatory reasons, Australia, Canada, the European Union, Japan and the United States maintain publicly available databases of field trials of GM plant varieties. The field test data provide information on both the amount of development work to produce new GM plant varieties and the types of GM traits under development (see the methodological note)
- Field trials can be undertaken for one genetic trait or for two or more traits simultaneously (stacked traits). The results presented here count each trait, rather than each field trial

- Field trial data are not available for Japan after 2001, and are therefore not included here. The most comparable recent time period for the remaining countries is between 2002 and 2004. The field trials are aggregated over these three years because of the low number of trials in the European Union and Australia
- Between 2002 and 2004 inclusive, 4,970 traits were field tested in the United States, Canada, the European Union, and Australia combined. The United States accounted for 70.5% of the total, Canada for 23.1%, Europe for 4.8% and Australia for 1.5%
- Canada has the highest intensity for field trials of specific traits, with 35.1 traits field tested per billion PPP$ of agricultural output at producer prices and 36.2 traits field tested per million population. The United States is second, with 18.2 traits field tested per billion PPP$ of output and 12.0 traits field tested per million population
- The field trials cover five major categories of genetic traits: tolerance to commercial herbicides, pest resistance to viruses, bacteria, insects, nematodes and fungi; improvements to product quality such as the type of oils, starches, sugars or cellulose in the plant, and improvements to agronomic characteristics such as yield or salt and cold tolerance. A fifth type of field trial covers technical traits, such as marker genes, genetic containment or other triats that are unclassifiable
- A large percentage of the field trials in the 1990s concerned herbicide tolerance. Other types of GM technology offer potential environmental advantages by reducing pesticide use and improving agronomic

characteristics (which could reduce fertilizer use). In addition, 'second generation' traits for improved product quality, such as low phytase animal feeds or improved oil seeds, could increase the value added of crop production, replace non renewable industrial feed stocks with renewable agricultural feed stocks, and provide a range of environmental benefits

- An increase over time in the share of product quality traits out of all traits is a marker for increasing interest in the commercial possibilities of second generation traits, which could reach the market within two to six years after the field trials. Trend data are available for the United States, Canada, and the EU-15
- Over time, the share of domestic field tests for product quality fell in the United States from 24% in 1995 to 12% in 2000, but increased after 2004, reaching 24% of American trials in 2005. The percentage of trials for product quality traits in the European Union reached 28% in 2003, but this is based on a total of only 111 traits that were field tested, compared to 814 traits tested in field trials in the United States in 2003. There is no consistent trend for Canada in the share of product quality traits
- The share of all domestic field trials for agronomic traits has increased the most in Canada, from 4% of all Canadian field trials of traits in 1998 to 24% of Canadian trials in 2004, followed by a slight decline to 22% in 2005. There is also a moderate upward trend in the United States since 2000, increasing from 7% to 17% of American field trials for agronomic traits in 2005.
- The United States accounted for over 50% of all

field trials for each trait category between 2002 and 2004 inclusive. The United States conducted 81.4% of the 795 tests for product quality traits between 2002 and 2004 in the United States, Canada, the European Union, and Australia combined. The United States also conducted 86.3% of all trials during these three years for pest resistance traits and 54.9% of all trials for agronomic traits.

Methodology

In the United States, the European Union, Canada and Australia field trials of new GM plant varieties are registered and the data are publicly available. Field trials cover a comparatively late stage of the development of GM varieties, as they do not include greenhouse and laboratory trials. Consequently, field trials provide evidence on relatively late stage research that could be ready for commercialisation within two to six years

Field trial data have many of the advantages and limitations of patents. Both provide a measure of investment in particular lines of research by firms and public sector institutions to develop new plant varieties (field trials) or inventions (patents), but in both cases there is no direct relationship between the number of trials or patents and the outcome in terms of commercialised GM varieties or inventions. A series of trials can be abandoned, with no commercialisation of the GM variety, and there is a large range in the number of field trials required to develop a GM variety. For example, several hundred field trials were conducted in the United States to alter the ripening characteristics of a tomato variety whereas only 15 trials were required to develop a virus resistant papaya variety

In the United States, field tests of GM varieties that have already received approval do not need to be registered, the bioeconomy. The bioeconomy, a new concept, covers a broad range of economic activities, each benefiting from new discoveries, related products and services arising out of the biosciences. The project will assess how pervasive biotechnological applications are likely to become, the prospects for further development over the next two to three decades and the potential impact on the economy and society.

(iii) *OECD Biotechnology Statistics 2006: 22-May-2006*

Just released, the 2006 edition OECD Biotechnology Statistics includes data for 2,3 OECD countries and 2 observer countries, plus China (Shanghai), and takes a major step forward in improving the comparability of biotechnology indicators among countries.

(iv) *ICGB Biotechnology Newsletter No. 16 Published: 18-May-2006*

The purpose of the OECD Biotechnology Update is to provide up-to-date information on the diverse activities at OECD related to biotechnology. The newsletter is mainly intended for delegates to OECD meetings who are familiar with OECD's work. However, we hope that it also serves a purpose to the wider biotech community.

(v) *Three New Biosafety Consensus Documents Published: 09-May-2006*

The Consensus Document on the Biology of Capsicum Annuum Complex (chili peppers, hot peppers and sweet peppers), The Consensus Document on Information Used in the Assessment of Environmental Applications involving Acidithiobacillus and The Consensus Document on the Biology of Western White Pine (Pinus monticola Dougl. ex D. Don) have been published.

(vi) *Innovation in Pharmaceutical Biotechnology: Comparing National Innovation Systems at the Sectoral Level: 07-Apr-2006*

What are the specific characteristics of innovation in pharmaceutical biotechnology? How do biopharmaceutical innovation systems in OECD countries perform and which policies are suitable to foster performance? This publication examines the innovation system in pharmaceutical biotechnology in eight OECD countries. Based on rich evidence, it draws policy recommendations to foster innovation in biopharmaceuticals advocating an integrated policy approach.

(vii) *Points to Consider for Consensus Documents on the Biology of Cultivated Plants: 31-Mar-2006*

The Points to Consider document, a structured explanatory checklist of relevant points to consider in preparing or evaluating a consensus document on the biology of cultivated vascular plant species, is now available.

(viii) *Publication of Two Consensus Documents: 05-Dec-2005*

The Consensus Document on the Biology of Papaya (*Carica Papaya*) and the Consensus Document on the Biology of *Pleurotus spp*. (Oyster Mushroom) have been published.

(ix) *Just Published: Health Technologies and Decision Making: 28-Oct-2005*

Health technology has great potential to change our understanding of disease, transform delivery of health-care services, and improve health outcomes.

This book examines how countries can successfully manage the opportunities and challenges arising from health-related technology by optimising decision-making processes, recognising the value of innovation, dealing with uncertainty, and producing and co-ordinating health technology assessment. It also considers health systems' capacity to respond to the particular challenges of fast-developing health-related biotechnologies.

Molecular Modeling

Proteins are fundamental components of all living cells. They exhibit an enormous amount of chemical and structural diversity, enabling them to carry out an extraordinarily diverse range of biological functions. Proteins help us digest our food, fight infections, control body chemistry, and in general, keep our bodies functioning smoothly. Scientists know that the critical feature of a protein is its ability to adopt the right shape for carrying out a particular function. But sometimes a protein twists into the wrong shape or has a missing part, preventing it from doing its job. Many diseases, such as Alzheimer's and "mad cow", are now known to result from proteins that have adopted an incorrect structure. Identifying a protein's shape, or structure, is key to understanding its biological function and its role in health and disease. Illuminating a protein's structure also paves the way for the development of new agents and devices to treat a disease. Yet solving the structure of a protein is no easy feat. It often takes scientists working in the laboratory months, sometimes years, to experimentally determine a single structure. Therefore, scientists have begun to turn toward computers to help predict the structure of a protein based on its sequence. The challenge lies in developing methods for accurately and reliably understanding this intricate relationship.

Allosteric Proteins

Under certain conditions, a protein may have a stable alternate conformation, or shape, that enables it to carry out a different biological function. Proteins that exhibit this characteristic are called allosteric. The interaction of an allosteric protein with a specific cofactor, or with another protein, may influence the transition of the protein between shapes. In addition, any change in conformation brought about by an interaction at one site may lead to an alteration in the structure, and thus function, at another site. One should bear in mind, though, that this type of transition affects only the protein's shape, not the primary amino acid sequence. Allosteric proteins play an important role in both metabolic and genetic regulation. Traditionally, a protein's structure was determined using one of two techniques: X-ray crystallography or nuclear magnetic resonance (NMR) spectroscopy.

X-ray Crystallography

Crystals are a solid form of a substance in which the component molecules are present in an ordered array called a lattice. The basic building block of a crystal is called a unit cell. Each unit cell contains exactly one unique set of the crystal's components, the smallest possible set that is fully representative of the crystal. Crystals of a complex molecule, like a protein, produce a complex pattern of X-ray diffraction, or scattering of X-rays. When the crystal is placed in an X-ray beam, all of the unit cells present the same face to the beam; therefore, many molecules are in the same orientation with respect to the incoming X-rays. The X-ray beam enters the crystal and a number of smaller beams emerge: each one in a different direction, each one with a different intensity. If an X-ray detector, such as a piece of film, is placed on the opposite side of the crystal from the X-ray source, each diffracted ray, called a reflection, recorded in 2003 for five countries: Slovak Republic, Poland, Hungary, Greece and Austria

- Firms have access to many other sources of funds than venture capital, including government investment, the stock market, banks, and private investors. Survey data on the venture capital share of the total amount of capital raised by biotechnology firms in a specific year are available for Israel, Denmark and Canada and for Belgium for the number of firms. The venture capital share was 6% in Israel in 2002, 13% in Denmark in 2002, and 22% in Canada in 2003. In Belgium, 16% of 49 biotechnology firms reported raising venture capital funds in 2003

Biotechnology is having an increasingly important impact on a number of sectors and disciplines. OECD has been working on biotechnology-related topics for over 25 years. These include scientific, industrial, health and agricultural applications. It also has a strong record on safety issues in biotechnology.

(i) *OECD Invites Public Comment on Draft guidelines for quality assurance in molecular genetic testing: 13-Jul-2006*

These guidelines are meant to encourage quality assurance systems for human genetic and genomic testing, to guarantee the international exchange of clinical samples and access to rare disease testing, and to help meet OECD member countries' efforts to supply high quality healthcare to their citizens.

(ii) *The Bioeconomy to 2030: Designing a Policy Agenda: 21-Jun-2006*

In 2006, the International Futures Programme (IFP) began an 18-month project that aims to draft a broad policy agenda for governments in respect to are in English. The most recent data are from the National Science Foundation's 2006 Science and Engineering Indicators

- Whether or not an alliance is made public and subject to a newspaper report will depend on the interests of the partners and the importance of the alliance to readers. Therefore, the CATIMERIT database is likely to exclude small alliances and those that the partners do not wish to publicly disclose. In addition, the database favours publications in English and consequently alliances from English-speaking countries such as the United States are likely to be over represented
- Results are only available for major countries or regions: the United States, Europe, Japan, and non-triad countries (involving a country outside the previous three)
- The share of all CATI-MERIT alliances that involve biotechnology has been increasing over time, from 11% of the total in 1990 to 53% in 2003
- Between 2001 and 2003 inclusive, 1,055 reported biotechnology alliances were included in the CATI-MERIT database. An alliance can include firms from two or more of the four countries or regions, or it can only include domestic firms. A partner from the United States was involved in 72.7% of the 1,055 biotechnology alliances over these three years, a European partner in 49.7%, a Japanese partner in 8.2%, and a partner from a non-triad country in 14.8%
- The total number of biotechnology alliances increased from 45 in 1990 to 368 in 2003. The growth in alliances was greatest for those involving partners

which decreases the comparability between Europe and the United States. The UNU-MERIT field trial database used here includes American data for both releases and notifications (an expedited type of release permit). For all countries, the UNU-MERIT database excludes non-plant field tests. The original field test data for Europe are available from http://biotech.jrc.it/deliberate/gmo.asp, for the United States from http://www.aphis.usda.gov, for Canada from http://www.inspection.gc.ca, and for Australia from http://www.ogtr.gov.au

The United States provides ten identifiers for the purpose of each trait. These identifiers were used by UNU-MERIT to identify field trials of specific traits for herbicide tolerance, pest resistance, product quality, agronomic characteristics, and other types of traits. The European Union, Canada, and Australia provide information on the trait but do not include an identifier. UNU-MERIT used the data from the United States and other sources to assign each trait in these countries to one of the five main categories. This classification system contains an unknown but small amount of error because some genetic traits can be used for different purposes. In a small number of trials insufficient detail is provided to accurately determine the purpose of a trial. These are assigned to the 'other' category. All results are based on field tests of specific traits. Approximately 20% of the field tests are for two or more 'stacked' traits. In these cases, each trait is counted separately

Biotechnology Alliance

- The CATI-MERIT database collects information on strategic alliances by domestic and multinational firms for technology transfer or joint research in biotechnology from announcements or articles in newspapers and professional journals, many of which will produce a spot on the film. However, because only a few reflections can be detected with any one orientation of the crystal, an important component of any X-ray diffraction instrument is a device for accurately setting and changing the orientation of the crystal. The set of diffracted, emerging beams contains information about the underlying crystal structure. If we could use light instead of X-rays, we could set up a system of lenses to recombine the beams emerging from the crystal and thus bring into focus an enlarged image of the unit cell and the molecules therein. But the molecules do not diffract visible light, and X-rays, unlike light, cannot be focused with lenses. However, the scientific laws that lenses obey are well understood, and it is possible to calculate the molecular image with a computer. In effect, the computer mimics the action of a lens. The major drawback associated with this technique is that crystallization of the proteins is a difficult task. Crystals are formed by slowly precipitating proteins under conditions that maintain their native conformation or structure. These exact conditions can only be discovered by repeated trials that entail varying certain experimental conditions, one at a time. This is a very time consuming and tedious process. In some cases, the task of crystallizing a protein borders on the impossible.

Nuclear Magnetic Resonance (NMR) Spectroscopy

The basic phenomenon of NMR spectroscopy was discovered in 1945. In this technique, a sample is immersed in a magnetic field and bombarded with radio waves. These radio waves encourage the nuclei of the molecule to resonate, or spin. As the positively charged nucleus spins, the moving charge creates what is called a magnetic moment. The thermal motion of the molecule—the movement of the molecule associated with the temperature of the material—further creates a torque, or twisting force, that makes the magnetic moment "wobble" like a child's top. When the radio waves hit the spinning nuclei, they tilt even more, sometimes flipping over. These resonating nuclei emit a unique signal that is then picked up on a special radio receiver and translated using a decoder. This decoder is called the Fourier Transform algorithm, a complex equation that translates the language of the nuclei into something a scientist can understand. By measuring the frequencies at which different nuclei flip, scientists can determine molecular structure, as well as many other interesting properties of the molecule. In the past 10 years, NMR has proven to be a powerful alternative to X-ray crystallography for the determination of molecular structure. NMR has the advantage over crystallographic techniques in that experiments are performed in solution as opposed to a crystal lattice. However, the principles that make NMR possible tend to make this technique very time consuming and limit the application to small- and medium-sized molecules.

Computational Modeling

Researchers have been working for decades to develop procedures for predicting protein structure that are not so time consuming and that are not hindered by size and solubility constraints. To do this, researchers have turned to computers for help in predicting protein structure from gene sequences, a concept called homology modeling. The complete genomes of various organisms, including humans, have now been decoded and allow researchers to approach this goal in a logical and organized fashion. Before we go any further, it is important to define some common terminology used in this field.

Levels of Protein Structure

To produce proteins, cellular structures called ribosomes join together long chains of subunits. A set of 20 different subunits, called amino acids, can be arranged in any order to form a polypeptide that can be thousands of amino acids long. These chains can then loop about each other, or fold, in a variety of ways, but only one of these ways allows a protein to function properly. The critical feature of a protein is its ability to fold into a conformation that creates structural features, such as surface grooves, ridges, and pockets, which allow it to fulfill its role in a cell. A protein's conformation is usually described in terms of levels of structure. Traditionally, proteins are looked upon as having four distinct levels of structure, with each level of structure dependent on the one below it. In some proteins, functional diversity may be further amplified by the addition of new chemical groups after synthesis is complete.

The stringing together of thc amino acid chain to form a polypeptide is referred to as the primary structure. The secondary structure is generated by the folding of the primary sequence and refers to the path that the polypeptide backbone of the protein follows in space. Certain types of secondary structures are relatively common. Two well-described secondary structures are the alpha helix and the beta sheet. In the first case, certain types of bonding between groups located on the same polypeptide chain cause the backbone to twist into a helix, most often in a form known as the alpha helix. Beta sheets are formed when a polypeptide chain bonds with another chain that is running in the opposite direction. Beta sheets may also be formed between two sections of a single polypeptide chain that is arranged such that adjacent regions are in reverse orientation. The tertiary structure describes the organization in three dimensions of all of the atoms in the polypeptide. If a protein consists of only one polypeptide chain, this level then describes the complete structure. Multimeric proteins, or proteins that consist of more than one polypeptide chain, require a higher level of organization. The quaternary structure defines the conformation assumed by a multimeric protein. In this case, the individual polypeptide chains that make up a multimeric protein are often referred to as the protein subunits. The four levels of protein structure are hierarchal, that is, each level of the build process is dependent upon the one below it.

A protein's primary amino acid sequence is crucial in determining its final structure. In some cases, amino acid sequence is the sole determinant, whereas in other cases, additional interactions may be required before a protein can attain its final conformation. For example, some proteins require the presence of a cofactor, or a second molecule that is part of the active protein, before it can attain its final conformation. Multimeric proteins often require one or more subunits to be present for another subunit to adopt the proper higher order structure. In any case, as we stated earlier, the entire process is cooperative, that is, the formation of one region of secondary structure determines the formation of the next region.

Some Basic Theory

It is theorized that proteins that share a similar sequence generally share the same basic structure. Therefore, by experimentally determining the structure for one member of a protein family, called a target, researchers have a model on which to base the structure of other proteins within that family. Moving a step further, by selecting a target from each superfamily, researchers can study the universe of protein folds in a systematic fashion and outline a set of sequences associated with each folding motif. Many of these sequences may not demonstrate a resemblance to one another, but their identification and assignment to a particular fold is essential for predicting future protein structures using homology modeling. The scientific basis for these theories is that a strong conservation of protein three-dimensional shape across large evolutionary distances—both within single species, between species, and in spite of sequence variation—has been demonstrated again and again. Although most scientists choose high-priority structures as their targets, this theory provides the option to choose any one of the proteins within a family as the target, rather than trying to achieve experimental results using a protein that is particularly difficult to work with using crystallographic or NMR techniques.

Maximizing Results

Specific tasks must be carried out to maximize results when determining protein structure using homology modeling. First, protein sequences must be organized in terms of families, preferably in a searchable database, and a target must be selected. Protein families can be identified and organized by comparing protein sequences derived from completely sequenced genomes. Targets may be selected for families that do not exhibit apparent sequence homology to proteins with a known three-dimensional structure. Next, researchers must generate a purified protein for analysis of the chosen target and then experimentally determine the target's structure, either by X-ray crystallography and/or NMR. Target structures determined experimentally may then be further analyzed to evaluate their similarity to other known protein structures and to determine possible evolutionary relationships that are not identifiable from protein sequence alone. The target structure will also serve as a detailed model for determining the structure of other proteins within that family. In favorable cases, just knowing the structure of a particular protein may also provide considerable insight into its possible function.

Molecular Modeling Database

NCBI's Molecular Modeling DataBase (MMDB), an integral part of our Entrez information retrieval system, is a compilation of all of the PDB three-dimensional structures of biomolecules. The difference between the two databases is that the MMDB records reorganize and validate the information stored in the database in a way that enables cross-referencing between the chemistry and the three-dimensional structure of macromolecules. By integrating chemical, sequence, and structure information, MMDB is designed to serve as a resource for structure-based homology modeling and protein structure prediction

NCBI has also developed a three-dimensional structure viewer, called Cn3D, for easy interactive visualization of molecular structures from Entrez. Cn3D serves as a visualization tool for sequences and sequence alignments. What sets Cn3D apart from other software is its ability to correlate structure and sequence information. For example, using Cn3D, a scientist can quickly locate the residues in a crystal structure that correspond to known disease mutations or conserved active site residues from a family of sequence homologs, or sequences that share a common ancestor. Cn3D displays structure-structure alignments along with the corresponding structure-based sequence alignments to emphasize those regions within a group of related proteins that are most conserved in structure and sequence. Cn3D also features custom labeling options, high-quality graphics, and a variety of file exports that together make Cn3D a powerful tool for literature annotation.

Protein Data Bank

The PDB was the first "bioinformatics" database ever built and is designed to store complex three-dimensional data. The PDB was originally developed and housed at the

Brookhaven National Laboratories but is now managed and maintained by the Research Collaboratory for Structural Bioinformatics (RCSB). The PDB is a collection of all publicly available three-dimensional structures of proteins, nucleic acids, carbohydrates, and a variety of other complexes experimentally determined by X-ray crystallography and NMR.

PDBeast: Taxonomy in MMDB

Taxonomy is the scientific discipline that seeks to catalog and reconstruct the evolutionary history of life on earth. NCBI's Structure Group, in collaboration with NCBI's taxonomists, has undertaken taxonomy annotation for the structure data stored in MMDB. A semi-automated approach has been implemented in which a human expert checks, corrects, and validates automatic taxonomic assignments. The PDBeast software tool was developed by NCBI for this purpose. It pulls text-descriptions of "Source Organisms" from either the original PDB-Entries or user-specified information and looks for matches in the NCBI Taxonomy database to record taxonomy assignments.

Structure Similarity Searching Using VAST

As just noted, a sequence-sequence similarity program provides an alignment of two sets of sequences. A structure-structure similarity program provides a three-dimensional structure superposition. Structure similarity search services are based on the premise that some measure can be computed between two structures to assess their similarities, much the same way a BLAST alignment is predicted. VAST or the Vector Alignment Search Tool, is a computer algorithm developed at NCBI for use in identifying similar three-dimensional protein structures. VAST is capable of detecting structural similarities between proteins stored in MMDB, even when no sequence similarity is detected. VAST Search is NCBI's structure-structure similarity search service that compares three-dimensional coordinates of newly determined protein structures to those in the MMDB or PDB databases. VAST Search creates a list of structure neighbors, or related structures, that a user can then browse interactively. VAST Search will retrieve almost all structures with an identical three-dimensional fold, although it may occasionally miss a few structures or report chance similarities.

Detecting New Sequence Similarities: BLAST against MMDB

Comparison, whether of structural features or protein sequences, lies at the heart of biology. The introduction of BLAST, or The Basic Local Alignment Search Tool, in 1990 made it easier to rapidly scan huge databases for overt homologies, or sequence similarities, and to statistically evaluate the resulting matches. BLAST works by comparing a user's unknown sequence against the database of all known sequences to determine likely matches. Sequence similarities found by BLAST have been critical in several gene discoveries. Hundreds of major sequencing centers and research institutions around the country use this software to transmit a query sequence from their local computer to a BLAST server at the NCBI via the Internet. In a matter of seconds, the BLAST server compares the user's sequence with up to a million known sequences and determines the closest matches. Not all significant homologies are readily and easily detected, however. Some of the most interesting are subtle similarities that do not always rise to statistical significance during a standard BLAST search. Therefore, NCBI has extended the statistical methodology used in the original BLAST to address the problem of detecting weak, yet significant, sequence similarities. PSI-BLAST, or Position-Specific Iterated BLAST, searches sequence databases with a profile constructed using BLAST alignments, from which it then constructs what is called a position-specific score matrix. For protein analysis, the new Pattern Hit Initiated BLAST, or PHI-BLAST, serves to complement the profile-based searching that was previously introduced with PSI-BLAST. PHI-BLAST further incorporates hypotheses as to the biological function of a query sequence and restricts the analysis to a set of protein sequences that is already known to contain a specific pattern or motif.

Application to Biomedicine

Although the information derived from modeling studies is primarily about molecular function, protein structure data also provide a wealth of information on mechanisms linked to the function and the evolutionary history of and relationships between macromolecules. NCBI's goals in adding structure data to its Web site are to make this information easily accessible to the biomedical community worldwide and to facilitate comparative analysis involving three-dimensional structure.

Conserved Domain Architecture Retrieval Tool

NCBI's Conserved Domain Architecture Retrieval Tool (CDART) displays the functional domains that make up a protein and lists other proteins with similar domain architectures. CDART determines the domain architecture of a query protein sequence by comparing it to the CDD, a database of conserved domain alignments, using RPS-BLAST. The Conserved Domain Architecture Retrieval Tool then compares the protein's domain architecture to that of other proteins in NCBI's non-redundant sequence database. Related sequences are identified as those proteins that share one or more similar domains. CDART displays these sequences using a graphical summary that depicts the types and locations of domains identified within each sequence. Links to the individual sequences as well as to further information on their domain architectures are also provided. Because protein domains may be considered elementary units of molecular function and proteins related by domain architecture may play similar roles in cellular processes, CDART serves as a useful tool in comparative sequence analysis.

Conserved Domain Database

The Conserved Domain Database (CDD) is a collection of sequence alignments and profiles representing protein domains conserved in molecular evolution. It includes domains from SMART and Pfam, two popular Web-based tools for studying sequence domains, as well as domains contributed by NCBI researchers. CD Search, another NCBI search service, can be used to identify conserved domains in a protein query sequence. CD-Search uses RPS-BLAST to compare a query sequence against specific matrices that have been prepared from conserved domain alignments present in CDD. Alignments are also mapped to known three-dimensional structures and can be displayed using Cn3D (see above).

COGs: Phylogenetic Classification of Proteins

The database of Clusters of Orthologous Groups of proteins (COGs) represents an attempt at the phylogenetic classification of proteins— a scheme that indicates the evolutionary relationships between organisms—from complete genomes. Each COG includes proteins that are thought to be orthologous. Orthologs are genes in different species derived from a common ancestor and carried on through evolution. COGs may be used to detect similarities and differences between species for identifying protein families and predicting new protein functions and to point to potential drug targets in disease-causing species. The database is accompanied by the COGnitor program, which assigns new proteins, typically from newly sequenced genomes, to pre-existing COGs. A Web page containing additional structural and functional information is now associated with each COG. These hyperlinked information pages include: systematic classification of the COG members under the different classification systems; indications of which COG members (if any) have been characterized genetically and biochemically; information on the domain architecture of the proteins constituting the COG and the three-dimensional structure of the domains if known or predictable; a succinct summary of the common structural and functional features of the COG members as well as peculiarities of individual members; and key references.

Gas Sensors

MS Modeling's DMol^3 has been used to investigate the nitrogen dioxide sensing mechanism of tin dioxide nanoribbons. Understanding the sensing mechanism will enable the efficient design and manufacture of nanoscale chemical sensors—an important commercial application of nanotechnology. Practical challenges with carbon nanotubes involving cost of synthesis, control of chirality and diameter, separation from bundles, and attachment of functional groups have prompted researchers to explore other types of one-dimensional nanostructures. Two of the most promising candidates in this regard are nanowires and nanoribbons.

As the name suggests, nanoribbons are solid objects (unlike nanotubes, which are hollow) with a near-uniform rectangular cross-section. So far, nanoribbons have primarily been synthesized from the oxides of metals and semiconductors. In particular, SnO_2 and ZnO nanoribbons have been materials systems of great current interest because of potential applications as catalysts, in optoelectronic devices, and as chemical sensors for pollutant gas species and biomolecules. Although they grow to tens of microns long, the nanoribbons are remarkably single-crystalline and essentially free of dislocations. Thus they provide an ideal model for the systematic study of electrical, thermal, optical, and transport processes in one-dimensional semiconducting nanostructures, and their response to various external process conditions. Recent experiments with SnO_2 nanoribbons indicate that these are highly effective in detecting even very small amounts of harmful gases like NO_2. Upon adsorption of these gases, the electrical conductance of the sample decreases by more than an order of magnitude. More interestingly, it is possible to get rid of the adsorbates by shining UV light, and the electrical conductance is completely restored to its original value. Such single-crystalline sensing elements have several advantages over conventional thin-film oxide sensors: low operating temperatures, no ill-defined coarse grain boundaries, and high active surface-to-volume ratio. To be able to fully commercialize their potential, it is important to better understand the sensing mechanism of such systems. With this goal in mind, researchers at the Brookhaven National Laboratory, Lawrence Berkeley National Laboratory, and Accelrys used MS Modeling's $DMol^3$ to investigate the nitrogen dioxide sensing mechanism of SnO_2 nanoribbons. Reporting in *Nano Letters*, the researchers examined the NO_2-sensing mechanism of SnO_2 nanoribbons with exposed (1 0'1) and (0 1 0) surfaces. The density functional theory (DFT) calculations revealed that:

- The most stable adsorbed species involved an unexpected NO_3 group doubly bonded to Sn centers
- An orders-of-magnitude drop in electrical conductance can be explained by significant electron transfer to the adatoms
- Computed binding energies were consistent with

adsorbate stability up to 700 K, with X-ray absorption spectroscopy indicating predominantly NO_3 species on the nanoribbon surface.

The ability of the nanoribbons to sense O_2 and CO was also investigated. In the case of O_2, the response of the nanoribbon was highly sensitive to the concentration of O-vacancies on the surface. Thus, in the absence of any surface vacancies, the calculation predicted negligible charge transfer. However, when surface vacancies were present, an O_2 molecule can adsorb as a peroxide bridge, and withdraw a significant amount of electronic charge from the nanoribbon surface, thereby decreasing its electrical conductancc. In the case of CO adsorption, there was a net electron transfer from the CO to the nanoribbon surface. Thus the calculation predicted an increase in nanoribbon electrical conductance upon CO-adsorption, in agreement with experimental results. Researchers from the Department of Energy's Sandia National Laboratories and the University of New Mexico (UNM) have developed a new way of mimicking photosynthetic proteins to manipulate platinum at the nanoscale. The method has the potential of changing the metal's properties and benefiting emerging technologies. While we are in the early stages of research, we see the possibility of manipulating the nanoscale structure of platinum so that we can have control over the size, porosity, composition, surface species, solubility, stability and other functional properties of these metal nanostructures," says John Shelnutt, the Sandia scientist leading the research effort. "Such control means that the redesigned platinum could be used in many new applications, including catalysis, sensors and optoelectronic and magnetic devices." He adds that while research groups have reported a few platinum nanostructures — including nanoparticles, nanowires, nanosheets and others — the addition of new types of nanostructures is "highly desirable and potentially technologically important." Working with Shelnutt in the research are Frank van Swol from Sandia, UNM graduate student Yujiang Song and Eulalia Pereira from the University of Porto in Portugal. The new method of manipulating platinum was detailed in a paper in the *Journal of American Chemical Society* published in December. The idea for the technique is similar to photosynthesis, in which plants use the energy from sunlight to produce sugar. But instead of manufacturing sugar, the new method changes a platinum ion to the neutral metal atoms. The photosynthetic protein mimicks this repeatedly, allowing metal to be deposited as desired at the nanoscale.

The method involves putting porphyrins — the active part of photosynthetic proteins — along with the platinum salt in an aqueous solution of ascorbic acid at room temperature. The porphyrins are placed in specific locations in the solution where it is intended that metal should be deposited. For example, the porphyrins may be confined to micelles, or liposomes. Micelles are spherical assemblies of detergent molecules in which the heads are exposed to the water and the tails stick together in the interior. Liposomes are similar structures, but they are larger and have water on the inside and outside separated by a closed membrane — sort of like a cell. The membrane is composed of two layers of detergent molecules, with the heads on the inner and outer surface facing the water and the tails forming the interior of the membrane. When light is shined on the porphyrins located in these detergent structures, the porphyrins excite, becoming catalysts for platinum reduction and deposition. As this occurs, the metal grows onto the surfaces of the surfactant structures as a thin sheet or in other ways. In the case of micelles, the platinum grows into balls that look like the common toy "Koosh" ball. The ball size can be controlled by the amount of porphyrins and platinum in the solution, the amount of light illuminating the solution, and the amount of time the light is on. For the metals platinum and palladium that form these nanostructures, it is enough for the porphyrin molecule to grow only a small metal "seed" particle composed of about 500 atoms. When it reaches this size, the seed starts to catalyze its own rapid growth (by oxidation of ascorbic acid), budding off arms in all directions and creating the Koosh-ball-like nanostructures. The porphyrin provides a convenient method of making these seeds at the location and time desired, leading to a uniform and selectable nanostructure size The platinum nanostructures take on a different form when they are prepared under different conditions. When the porphyrin is in a micelle, the platinum nanostructures produced look like Koosh balls. When the porphyrin is in the bilayer membrane of a liposome, the platinum grows in 2-nanometer thick sheet or platinum lace on the outer surface of the membrane, giving circular sheets — sort of like two-dimensional Koosh balls. Under solution conditions for which the liposomes aggregate, growth can occur along the interfaces of the liposomes to give platinum foamlike materials and foam nanoballs.

The type of nanostructure is mainly determined by the type of surfactant assembly upon which the platinum grows and the extent of growth from the individual seed nanoparticles. Since the porphyrin remains attached to the platinum nanostructure and active in the presence of light, it can also perform other functions besides growing itself. For example, when illuminated with light, the platinum nanostructure evolves hydrogen from water. This reaction is similar to one of interest to car manufacturers looking for new ways to build automobiles powered by hydrogen fuel cells. Shelnutt says that in addition to structuring the platinum, the process also happens very fast. A few minutes in light will create many seeds, which then grow into the mature nanostructures in tens of minutes. And the process is easy to do.

Scientific Application Semiconductor Superlattices

A brief review of semiconductor superlattices is presented—their definition, basics of theory, types of materials and methods used for their preparation, the most important and unusual electrical and optical properties, which have already found their practical applications. When alternating layers of two semiconductors with different band gaps are deposited and thickness of these layers lies between 1-10 nm (which corresponds approximately to 10–100 atomic layers), what we obtain is semiconductor superlattice. The mostly known example of such a structure is GaAs/AlAs system depicted schematically together with its corresponding potential profile.

Differences in band gaps of materials lead to discontinuities in valence and conduction band and a set of square potential wells (low band gap semiconductor) separated by potential barriers (high band gap semiconductor) forms. Let us consider one of these wells from the quantum mechanical point of view. If we assume that an electron mean free path within a well is substantially longer than well width and the barrier width is infinite, what we obtain is a classical problem of a particle confined in a square well of finite depth. Unfortunately, in the real-life systems we cannot obtain perfect separation of adjacent wells and this results in coupling of their wave functions. Thus, instead of discrete levels, this yields us subbands. An example of theoretical analysis of the GaAs/AlAs system shown, where an equal width, a, of well and barrier layer is assumed. Two main features common for other superlattice structures are easily seen when a is decreased. First, an energy distance between ground state and subsequent bound states increases, which happens because the well width is getting smaller. Second, the discrete levels split into the subbands as a result of thinner barrier layers and poorer separation of neighbouring wells. At this point, it is also important to notice, that the energy discontinuity, i.e. well depth, cannot be simply derived from such material constants as an energy gap, the conduction and valence band structure etc. in each semiconductor because of the lack of knowledge of the common energy reference. This parameter must be either calculated or measured. Historically, the crystalline superlattices where the first to produce, starting from pioneering work of Esaki and Tsu (1970). What is crucial in fabrication of such structures is nearly perfect match in their lattice constants. Otherwise the density of defects becomes so large that it results in strong electron scattering and obscures quantum phenomena.

This is why a molecular beam epitaxy technique is utilised in their deposition and the number of materials, which can be combined to create superlattices, is restricted. The best known and studied are GaAs/AlAs and GaAs/AsxGa1-xAs, InAs/GaSb are also worth mentioning. If interface defects cause such difficulties, why not to approach the problem from the opposite side—let us build the superlattice from the materials, where lattice constants no longer exist, i.e. amorphous semiconductors. This idea, apparently irrational, proved absolutely sensible in 1983, when Abeles and Tiedje reported in their paper fabrication of a-Si1-xNx:H/a-Si:H and a-Ge:H/a-Si:H systems. They stated that the strict requirements concerning lattice matching in crystalline superlattices can be relaxed due to amorphous structure by its very nature and a hydrogen content which plays an important role by passivating coordination defects. Since then, many other materials were combined to produce superlattices : a-Si:H/a-C:H, a-Si:H/a-SiN:H, a-Si:H/a-SiNx, a-Si:H/a-SiOx, a-Si:H/a-Ge:H, CdSe/SeTe, CdSe/Se, a-Si3N4:H/a-Si:H, a-C:H superlattices where the band gap is tailored by changing deposition conditions. An impressive, comparing to the crystalline case, number of types of amorphous superlattices results from the fact that in this case we have much more freedom in choosing components of our systems as we are not limited by the lattice matching condition.It is also important to emphasise that the techniques utilised to produce amorphous films - plasma deposition, thermal evaporation, sputtering—are relatively cheaper than molecular beam epitaxy and easier to implement in industrial applications.

On the other hand, in case of amorphous multilayers it is more difficult to obtain abrupt interfaces between subsequent layers. Different methods are employed : pulsed application of microwave plasma during continuos r.f. glow discharge in Si2H6/N2, by which N2 is decomposed and incorporated into the growing film, nitridation of a-Si:H to attain a-Si3N4:H, lowering the growth rate in order to avoid fabrication of transition layers when the deposition conditions change, quick exchange of source gases. However, sometimes smooth transition from one layer to another is intentionally introduced, which may lead to sinusoidal potential profile, instead of a step-like. For the sake of convenience, electrical properties, particularly conduction phenomena, may be considered in terms of what is happening when we study currents flowing in-plane (parallel to the layers) or perpendicular to them. As it has been stated above, quantum confinement causes shift of the first bound state allowed for charge carriers. This must of course result in the increase of an activation energy of conductivity, E_a. In general, the smaller well width, the higher E_a, thus in-plane electrical conductivity decreases as compared to the unlayered material.

This is because in amorphous semiconductors, unlike crystalline ones, there is a finite density of states within a mobility gap. When two such materials are put into contact, charge carriers may be transferred through an interface, building up accumulation and depletion layers, and as a result, the conductivity in one of the materials increases to such a degree that obscures quantum effects. A completely different picture appears when conductivity in the direction perpendicular to the layers is considered. If there is no electric field, the current does not flow of course. When the small field is applied, a tunneling current can flow through the barriers.

As the field is increased, an alignment between the energy subbands in adjacent wells is getting worse so observed maximum is a compromise between a driving force (electric field) and subbands mismatch. When the field is increased further, the subbands misalignment is so high, that charge transport becomes very inefficient and the current—field characteristics exhibits minimum. If the field becomes higher, we can have the situation, that the lower subband in one quantum well is aligned with the upper one in the adjacent well. Transport is efficient again.

The most important feature of this characteristics is a region of negative differential resistance where an increase of the electric field leads to a decrease of the electric current. More detailed theoretical considerations of this matter can be found elsewhere. This kind of behaviour has been observed in both crystalline and amorphous multilayers; however, in the crystalline case it is more distinct and resembling model behaviour. An example of characteristics for DLC superlattice is shown, where current-voltage profile together with its first and second differentials is presented. Bumps observed on the current-voltage curve are typical of resonant tunneling. The start of the negative differential resistance region at 3.1 V corresponds to the loss of subbands alignment. It should be noted that the conductivity in the direction perpendicular to the superlattice layers is by its nature very similar to what Esaki found in *p-n* germanium junctions and what is now applied in construction of tunnel diodes. Typical *blue shift* effect is shown in Figs. 20.7 and 20.8. The solid lines represents fit to a theoretical model of a one-dimensional quantum well with no coupling between energy levels of adjacent wells. We have discussed electrical and optical phenomena which characterise superlattice structures which found their explanation on the basis of quantum size effect.

However, alternative interpretations of experimental data do exist as well. Arsenault et al. claim that bumps observed in the current-voltage characteristics for perpendicular conductivity in amorphous multilayers, usually attributed to resonant tunneling can be explained in a different way. They suggest that at least one of these bumps is simply a transition between two transport mechanisms: space charge limited current (low electric field) and multiple hopping (high field). The most convincing justification of their hypothesis is that they observed the same bumps for both quantum well structures and single barriers.

Quantum mechanical explanation of the *blue shift* has been criticised in a number of research papers as well. In our discussion it has been assumed that described multilayers were periodic system, characterized, from the geometrical point of view, only by one well and one barrier width. However it does not have to be the only approach. Another and very original view is presented in some researches. Superlattices of these researches consist of the layers of a-

Ge and a-Si deposited using dual-target magnetron sputtering. The most interesting thing is their topology generated by Cantor sets. The modulation algorithm according to the so called "Cantor middle—third" set together with corresponding multilayer structure is presented.

4

Biotech Science: Tools and Techniques

Whitehead Institute for Biomedical Research Massachusetts Institute of Technology

In 1992, we celebrated the 500th anniversary of quite a remarkable event in human history — Christopher Columbus' discovery of the Americas. In the meeting between the two cultures, European and American, the Native Americans presented Columbus with a remarkable gift more valuable than the gold and spices that he set out to find. The gift to Columbus was a super cereal that the Native Americans developed.

When Columbus arrived in the New World in 1492, he may have observed scenes like this depicted by the great Mexican artist, Diego Rivera. You can see here a feverish marketplace but the most important thing in this scene — one every Native American would have seen immediately — is in this enlargement and that is corn. Corn was the staple of the New World and it was a New World invention. It was not found in Europe before Columbus' visit.

The immortality and potentially unlimited developmental capacity of human embryonic stem (ES) cells ignite the imagination. After months or years of growth in culture dishes, these cells retain the ability to form cell types ranging from heart muscle to nerve to blood—possibly any cell in the body. Because of their unique developmental potential, human ES cells have widespread implications for human developmental biology, drug discovery, drug testing, and transplantation medicine. Indeed, human ES cells promise an essentially unlimited supply of specific cell types for *in vitro* experimental studies and for transplantation therapies for diseases such as heart disease, Parkinson's disease, leukemia, and diabetes.The derivation of a human ES cell line destroys a human embryo. Thus, the derivation of human ES cells resurrected a fierce controversy over human embryo research in the United States, a controversy originally created by the development of human *in vitro* fertilization decades ago, but never completely resolved. In particular, the derivation of human ES cells led to a re-examination of the role of federal funding of human embryo research. In response to the intense public interest, President George W. Bush reviewed the potential of human ES cell research to improve the health of Americans.

In his national address on August 9, 2001, he stated "Federal dollars help attract the best and brightest scientists. They ensure new discoveries are widely shared at the largest number of research facilities, and the research is directed toward the greatest public good." On that basis, he directed federal funding to "explore the promise and potential of stem cell research," including, for the first time, human ES cell research. However, the President went on to restrict federal funding to research that used only those human ES cell lines derived prior to his address on August 9, 2001. This paper reviews progress in human ES cell research in the wake of that decision.

Genetic Modification of Human ES Cells

Although the human genome project is essentially

completed, we are ignorant about the function of most human genes.

Human ES cells provide a powerful new model for identifying the function of any human gene, and this requires efficient methods for genetic modification of human ES cells. Genetic manipulation of human ES cells is essential to elucidate gene function; direct the differentiation of ES cells to specific lineages; purify desired differentiated cell types from mixed populations of ES cell derivatives; use the differentiated derivatives of ES cells as a vehicle for gene therapy; and modulate the immune response to transplanted ES cell derivatives. Transfection methods routinely used for mouse ES cells generally fail to transfect human ES cells efficiently, but there have now been several approaches developed for human ES cells. Transient and stable integration of plasmids into human ES cells can be accomplished through specific transfection reagents, the best reagents yielding stable (drug-selectable) transfection rates of about 10 . Recently more labor-intensive, HIV-based, lentivirus vectors have been shown to transduce human ES cells at an efficiency rate of over 90%. This should allow complex mixtures of genes to be screened for specific phenotypic effects by a process termed "expression cloning". Homologous recombination allows the defined modifications of specific genes in living cells and has been used extensively with mouse ES cells. However, the differences between mouse and human ES cells delayed the development of homologous recombination in human ES cells. Except for viral approaches, high stable transfection efficiencies in human ES cells have been difficult to achieve, and in particular, electroporation protocols established for mouse ES cells do not work in human ES cells. Also, in contrast to mouse ES cells, human ES cells proliferate inefficiently from single cells, making screening procedures to identify rare homologous recombination events difficult. We have recently developed modified electroporation protocols to overcome these problems and have successfully targeted a ubiquitously expressed gene (*HPRT1)*, an ES cell-specific gene (*POU5F1)*, and a tissuespecific gene (Tyrosine hydroxylase: TH) in human ES cells.

The overall targeting frequencies for the three genes suggest that homologous recombination is a broadly applicable technique in human ES cells. Homologous recombination in human ES cells will be important for studying gene function *in vitro* and for lineage selection. For therapeutic applications in transplantation medicine, controlled modification of specific genes should be useful for purifying specific ES cell-derived, differentiated cell types from a mixed population; altering the antigenicity of ES cell derivatives; and giving cells new properties (such as viral resistance) to combat specific diseases. Homologous recombination in human ES cells might also be used for approaches combining therapeutic cloning with gene therapy. *In vitro* studies using homologous recombination in human ES cells will be particularly useful for learning more about the pathogenesis of diseases where mouse models have proven inadequate. For example, HPRT-deficient mice fail to demonstrate an abnormal phenotype, yet defects in this gene cause Lesch-Nyhan disease in children. *In vitro* neural differentiation of HPRT-deficient human ES cells or transplantation of ES cell-derived neural tissue to an animal model could help to understand the pathogenesis of Lesch-Nyhan syndrome.

Human ES Cells as a Model of Early Human Development

The excitement surrounding the prospective role of human embryonic stem (ES) cells in transplantation therapy has often overshadowed a potentially more important role as a basic research tool for understanding the development and function of human tissues. The use of human ES cells is particularly valuable to derive tissue for study that is difficult to obtain otherwise, and for which animal models are inadequate.

Human ES cells offer a new and unique window into the early events of human development, a period critical for understanding infertility, birth defects, and miscarriage. Because manipulation of the early post-implantation human embryo could jeopardize the health of the resulting child, it has never been possible to examine this important period of human development experimentally. Nearly all of what is known about early human development, especially in the early post-implantation period, is based on very rare histological sections of human embryos, or on an imperfect analogy to experimental studies in the mouse. The mouse has been the mainstay of mammalian experimental embryology because of its historical use, well-defined genetics, and favorable reproductive characteristics. However, early mouse development and early human development differ significantly. For example, human and mouse embryos differ in the expression of embryonic antigens; timing of embryonic genome expression; formation, structure, and function of the fetal membranes and placenta; and formation of an embryonic disc instead of an egg cylinder. Thus, if one is interested in the development of a human tissue known to differ significantly from the corresponding mouse tissue, such as the yolk sac or the placenta, studying a human model is desirable. The first differentiation event in mammalian embryos is the formation of the trophectoderm, the outer epithelial layer of the blastocyst. The trophectoderm is crucial for implantation of the embryo and gives rise to specialized populations of trophoblast cells in the placenta. Mouse and human placentas differ in structure and function, and these differences are clinically significant. For example, the placental hormone chorionic gonadotropin, which has an essential role in establishing and maintaining human pregnancy, is not even produced by the mouse placenta. When formed into chimeras with intact preimplantation embryos, mouse ES cells rarely contribute to the trophoblast, and the manipulation of external culture conditions has, to date, failed to direct mouse ES cells to form trophoblast. Spontaneously differentiated rhesus monkey or human ES cells do secrete modest amounts of chorionic gonadotropin, indicating the differentiation of trophoblast cells. Recently it was discovered that a single growth factor (BMP4) would induce human ES cells to differentiate to a pure population of early trophoblast. These early human trophoblast cells have never before been available for detailed study, and already this new experimental model has provided information about the specific genes that control the early development of the human placenta. The derivation of other early lineages from human ES cells *in vitro* to provide a more complete understanding of early human development is an active area of research.

Bibliography

Assady, S., *et al.*, "Insulin production by human embryonic stem cells," Diabetes 50: 1691-1697 (2001).

Chu, K., et *al.*, "Human neural stem cells can migrate, differentiate, and integrate after intravenous transplantation in adult rats with transient forebrain ischemia," Neuroscience Letters 343: 129-133 (2003).

Clark, J., "Squandering Our Technological Future," *The New York Times*, August 30, 2001, p. A19. See also the presentation of Thomas Okarma, President and CEO of Geron Corporation, before the Council on September 4, 2003.

Faraci, M., *et al.*, "Severe neurologic complications after hematopoietic stem cell transplantation in children," Neurology 59: 1895-1904 (2002).

Folkman, J. Therapeutic angiogenesis in ischemic limbs. Circulation 1998; 97:108- 110.

Gianessi, Leonard P. and Janet E. Carpenter, "Agricultural Biotechnology: Insect Control Benefits," National Center for Food and Agricultural Policy, 1999. Available on Biotechnology Industry Organization at www.bio.org.

Hogan, P., et al., "Economic Costs of Diabetes in the US in 2002," Diabetes Care 26: 917-932 (2003).

Kennedy M, Firpo M, Choi K, Wall C, Robertson S, Kabrun N, Keller G. A common precursor for primitive erythropoiesis and definitive haematopoiesis.

Kennedy, D., "Stem Cells: Still Here, Still Waiting," *Science* 300: 865 (2003).

Kerr, D. A., et al., "*Hum*an Embryonic Germ Cell Derivatives Facilitate Motor Recovery of Rats with Diffuse Motor Neuron Injury," The Journal of Neuroscience 23: 5131-5140 (2003).

Kerr, D. A., *et al.*, "Human Embryonic Germ Cell Derivatives Facilitate Motor Recovery of Rats with Diffuse Motor Neuron Injury," The Journal of Neuroscience 23: 5131-5140 (2003).

Labastie M-C, Cortes F, Romeo P-H, Dulac C, Peault B. Molecular identity of hematopoietic precursor cells emerging in the human embryo. Blood 1998;92:3624-3635.

Latham, S., "Ethics and Politics," *American Journal of Bioethics* 2(1): 46 (2002).

Lechner, A. and Habener, J.F., "Stem/progenitor cells derived from adult tissues: potential for the treatment of diabetes mellitus," American Journal of Physiology—Endocrinology and Me*tabolism* 284: E259-266 (2003).

Lewis, C. S. The Abolition of Man. New York: The Macmillan Company, 1947.

Liker, M., et al., "Human neural stem cell transplantation in the MPTP-lesioned mouse," Brain Research 971: 168-177 (2003).

Index

N

O

P

Q

R

S

T

U